Green Energy and Technology

Climate change, environmental impact and the limited natural resources urge scientific research and novel technical solutions. The monograph series Green Energy and Technology serves as a publishing platform for scientific and technological approaches to "green"—i.e. environmentally friendly and sustainable—technologies. While a focus lies on energy and power supply, it also covers "green" solutions in industrial engineering and engineering design. Green Energy and Technology addresses researchers, advanced students, technical consultants as well as decision makers in industries and politics. Hence, the level of presentation spans from instructional to highly technical.

Indexed in Scopus.

More information about this series at http://www.springer.com/series/8059

Angel Arcos-Vargas
Editor

The Role of the Electric Vehicle in the Energy Transition

A Multidimensional Approach

Springer

Editor
Angel Arcos-Vargas
Industrial Engineering and Management
University of Seville
Seville, Spain

ISSN 1865-3529 ISSN 1865-3537 (electronic)
Green Energy and Technology
ISBN 978-3-030-50632-2 ISBN 978-3-030-50633-9 (eBook)
https://doi.org/10.1007/978-3-030-50633-9

This Springer imprint is published by the registered company Springer Nature Switzerland AG
The registered company address is: Gewerbestrasse 11, 6330 Cham, Switzerland

Foreword

The automotive sector is a strategic industry and a major contributor to the Spanish economy. It is the first exporting sector and occupies an outstanding second position in manufacturing production, only surpassed by the agri-food sector. Spain is the second-largest producer of passenger cars in Europe (just behind Germany), the ninth in the world, and has been establishing itself for years as the leading European industrial vehicles manufacturer. These figures show the importance of a sector immersed in a deep transformation because of facing three important challenges.

The first of these challenges is the impact of digitization on each and every one of its dimensions. This transition has been so pervasive in manufacturing that it has been dubbed 'Industry 4.0', and it encompasses hyperconnectivity in production processes, additive manufacturing, robotics, cloud computing, and artificial intelligence. However, the continued digitalization of the entire value chain goes one step further in the automotive sector with connected and/or autonomous vehicle technologies. These technologies have the potential to change transportation on a global scale for they could improve safety, significantly alter transportation costs, and change traffic patterns and congestion. Likewise, the increasing shared mobility is possible by digitization. The Ministry of Industry, Trade and Tourism has launched a specific support programme to facilitate digital transformation, the 'Connected Industry 4.0' programme, which promotes the incorporation of digital knowledge in industrial companies. The programme offers support for advising companies on their digitization process and also aids for their investments.

A second important challenge for the automotive sector is the transition toward a decarbonized economy as well as a farther adoption of the circular economic model. Transport is key in Spain's decarbonisation roadmap since it leads to energy consumption in Spain, reaching nearly 42% of the final energy consumption. Only the passenger car segment represents 15% of the final energy consumed in our country. The huge dimension of the challenge demands a step change in both the breadth and scale of ambition. The associated benefits of bold and ambitious action to tackle transport emissions are significant: Improving citizen's health, driving sustainable economic growth and, in the end, creating better places to live. Our commitment toward an ecological transition and its opportunities for investment, innovation, and quality employment is articulated through different strategies that ensure a fair and orderly transition.

The 'National Energy and Climate Plan 2021–2030' (PNIEC) puts Spain on the path to achieve climate neutrality by 2050 and comply with the Paris Agreement. Specifically, this Plan considers that a fleet of 5.000.000 electric vehicles (such as passenger cars, vans, buses, and motorcycles) could be achieved in 2030. It exemplifies the government's commitment to electromobility. On the other hand, the Spanish 'Just Transition Strategy' focuses on those regions and people affected by the implementation of the aforementioned plan, providing a framework for action to optimize opportunities under the ecological transition.

In this regard, Electric Vehicles (EV) present a substantial technological shift, which require workers across the automotive supply chain to develop new skill sets. We are actively working to develop EV specific qualifications through continuing training to accompany businesses to repurpose ICE manufacturing facilities, to help workers develop new EV skills, as well as to ensure that independent dealerships and mechanics are equipped to advise buyers on EV options and to conduct repairs safely.

Far from being a disadvantage, such an ecological transition represents a unique industrial opportunity to drive innovation in the development of new zero or low emission propulsion systems and to continue leading our country as a benchmark in the manufacture of new alternative fuelled vehicle models. In that sense, the Ministry of Industry, Trade and Tourism has support programmes that offer soft loans to companies to achieve their investments in new production lines and vehicles.

Lastly, as a third challenge, the automotive sector has to face an international competition in a global market. It is obliged to innovate, renew itself and take advantage of all its potential to continue occupying its leading position in a sustainable way. The Spanish automotive industry has come a long way since the manufacture of the first EV in our country on an industrial scale was announced ten years ago. Since then, the commercialization of new alternative propulsion models by different brands has been increasing, and with it, the awards to the Spanish factories. This has allowed our plants to consolidate their competitive edge in a high-value niche, such as the alternative fuelled vehicles. Currently, Spain is projected to produce Battery Electric Vehicles (BEV) and Plug-in Hybrid Electric Vehicles (PHEV) in 2021 in nine out of our eleven factories that produce passenger cars and vans. These important advances are the result of the competitiveness of Spanish factories as well as the intense effort made by the Ministry I'm leading to create an attractive investment environment that encourages manufacturers to locate EV facilities in Spain, and to repurpose, rather than retire, existing ICE production lines for EVs. We must be ahead in the EV transition worldwide, if Spain is to retain its share in the current global market.

Our goal is to have prominent players throughout the entire value chain. Electromobility is not only the vehicle itself; components are critical element as well. We are working together with components companies to boost their competitiveness and to develop new solutions for the alternative fuelled, connected vehicles, especially in relation to batteries for which we address all phases of its value chain, from production to more nascent ones such as second life battery applications, battery disposal, and recycling. However, we are aware that drivers will not switch to EVs if they are not confident that they can charge their vehicle in local areas and on major roads. Spain has launched in recent years several aid programmes to stimulate the demand of EV vehicles and to facilitate the installation of charging infrastructure.

In conclusion, the automotive sector is undergoing a paradigm shift that must lead to a more sustainable, safe, connected, and intelligent mobility. The confluence of stricter environmental regulations, new consumer demands, and advances that push the boundaries of technology are transforming the vehicles profiles and their manufacturing. Spain is prepared to capture the industrial opportunities that this transformation is generating to drive innovation in the development of new zero or low emission value chains under the principle of technological neutrality.

In this journey, the Ministry of Industry, Trade and Tourism will continue to support the efforts of stakeholders to adapt to the technological challenges that mobility represents. It will be done with different instruments, among others, aid programmes, vehicle purchase incentives, public–private partnerships, interdisciplinary working groups to identify priority action areas, personalized advisory service on public funding or coordination across the policies and incentives set by different regions, cities, and the UE programmes.

We are working for placing Spain to seize the economic opportunities that will appear due to the multiple areas in which the vehicles of the future will play a role. The faster we act, the greater the benefits.

I would like to end these lines by thanking the authors for their contributions to this book. They provide clear insights to stakeholders in the electric vehicle transition and the size of the challenges we need to tackle.

Reyes Maroto Illera
Minister of Industry
Trade and Tourism—Government of Spain

Acknowledgements

There are many people who have contributed to the process of designing, documenting, and writing this work, whose first steps were taken four years ago, and there would be no room to thank them all, much less in the manner they merited.

In any case, I do think it is fair to thank the institutions that have always supported us, such as the Real Academia de Ingeniería of Spain, the journal Economía Industrial, the Ministry of Industry and Tourism of the Spanish Government, and the five universities in which the fifteen authors of this book are professors: Universidad de Sevilla, Universidad Politécnica de Madrid, Universidad Pontificia Comillas, Universidad de Deusto, and Universidad Autónoma de Chile.

Thank you all very much.

Introduction

Nowadays, nobody doubts that the future of mobility involves its electrification. Environmental concerns, the need to decarbonize the economy, the concentration of population in cities, the deterioration of productivity as a result of traffic congestion, technological progress in electrical storage systems, the improvement of communications with the introduction of 5G, the Internet of Things (IoT),… are some of the factors that are making it possible for this transition, which seemed like science fiction to us a decade ago, to become a reality.

The European Union's commitment to climate change is unequivocal, with all Member States having deposited instruments of confirmation to meet their pledges to reduce their national emissions by at least 40% between 1990 and 2030, setting an intermediate target of 20% by 2020 (COP-21, 2015). According to the latest projections from the States, there will be no problem in complying with the objectives for 2020 (in fact, they have already been achieved), but it will be difficult to meet the objective for 2040, with trend values of around 32%.

From the statements on the sources of emissions made by countries, it can be seen that most of them come from transport, so an orderly, efficient, and intelligent transition from conventional to electric transport would mean ensuring compliance with the planned reduction.

Many questions come to mind at this point: Are they really that expensive? Is the technology mature, are the targets ambitious? What measures are being incorporated by different countries? What are the differences between countries? Will the electricity system be able to supply this increase in demand? Are public policies adequate? What is the real contribution of electric vehicles to the environment? Will there be enough raw materials? What impact will it have on economic growth?… among others. In this book, we will try to clarify some of them, aiming to make them useful for users, companies, or regulators.

This book is structured in nine chapters and starts with a description of the different levels of presence of the electric vehicle in European countries, as well as a review of the different policies and fiscal measures adopted. Chapter 2 is dedicated to estimate the increase in electricity demand associated with the electrification of the electric vehicle; for this purpose, it develops a model based on the distance traveled and the typology of vehicle used (cars, buses, motorcycles,...) to finish by providing an estimate of the reduction in emissions associated with its total deployment. Once the demand is known, Chap. 3 reviews the different recharging technologies and estimates the possible impact that this new demand could have on the network.

In Chap. 4, an investigation on the determinants that explain the differences between countries is carried out, for which a stochastic frontier model is developed for data panel, enabling the identification of the significant factors for their introduction, as well as presenting a first evaluation and possible proposal for public policies. It arrives, as would be expected, that the existence of a public and fast recharging network is the determining factor in the growth of the number of electric vehicle registrations. Given that it is necessary to develop (and finance) such a charging network, the profitability that it could have for a private investor is analyzed in Chap. 5.

As far as the environmental impact is concerned, Chap. 6 studies the environmental aspects, for which a life cycle analysis is carried out. This analysis is then extended to the macroeconomic effects that can be foreseen, based on the input–output tables of each country.

One of the most common concerns of society is whether there will be enough raw materials to replace the existing assets, which is discussed in detail in Chap. 8. This work ends by analyzing the role that other non-conventional transport technologies might play in the transition scenario.

Although many other matters would remain to be explored, given the multidimensional nature of the case, an overview of the main dimensions of transport electrification in the economy has been provided throughout these nine chapters.

Angel Arcos-Vargas
Industrial Engineer, Ph.D., Economics, Ph.D.

Contents

EV International Landscape

José María Maza-Ortega, Francisco Jesús Matas-Díaz, and Ángel Arcos-Vargas

Abstract This chapter is devoted to give a general overview of the best practises in some European countries for maximizing the deployment of the electrical vehicle. After a brief introduction where the main reasons for promoting this technological transition are outlined, it has reviewed the general EU regulatory context. Then, different policies of relevant countries in the electrical vehicle development such as Denmark, France, Italy, Norway, Germany, Estonia, and The Netherland are analyzed. Finally, the different stakeholders involved in e-mobility are reviewed with a particular focus on the role than Distribution System Operators may have in the business model development.

1 Introduction

Too many countries are trying to foster the deployment of the electric vehicle (EV) with different actions and incentives to turn it into a reference element in the public and private transportation sector. The main reasons are diverse: economic benefits for the owners, development of new technology, increased energetic independence, reduced dependency on fossil fuels, etc. The major reason, however, is to achieve a drastic reduction of CO_2, NO_x, and particulate matter (PM) in the atmosphere.

Private vehicles are responsible for the 83% of the CO_2 emissions of the overall transportation sector. Particularly, they are the source of 80% of the total NO_x emissions and 60% of the emissions of particles (Agència de Salut Pública de Barcelona) [1]. This share increases in urban areas due to the current transport policies which promote the use of private transport means and the deployment of new roads leading to an intensive occupation of public space. The PM composition, continuously inhaled by citizens, may have a diverse mix which may be classified according to its size and behavior during their inhalation rather than its chemical formulation. Particles with aerodynamic diameter equal to or lower than 10 μm (PM10) usually pass throughout the throat. Those particles with aerodynamic diameter below 2.5 μm (PM2.5) may

J. M. Maza-Ortega · F. J. Matas-Díaz · Á. Arcos-Vargas (✉)
University of Seville, Seville, Spain
e-mail: aarcos@us.es

A. Arcos-Vargas (ed.), *The Role of the Electric Vehicle in the Energy Transition*, Green Energy and Technology, https://doi.org/10.1007/978-3-030-50633-9_1

reach the lungs. Finally, those ultrathin particles with a diameter below 0.1 μm may pass to the blood through the lung alveoli (Agència de Salut Pública de Barcelona [1]. Therefore, transportation sector can be considered as one of the main problems of public health in urban environments. According to two reports of the World Health Organization (WHO) published in 2018 [2, 3], more than a quarter of deaths of children below 5 years old are due to environmental contamination (1.7 million per year). An in-depth look reveals that 570 thousand deaths are due to respiratory infection and 270 thousand within the first month after the childbirth, a direct cause of it being the air contamination both in open and enclosed spaces [2]. A previous press note given the facts of a study carried out in 2014 revealed that 7 million people died as a consequence of the direct exposition to the atmospheric contamination in 2012 [4]. Moreover, a statistical analysis about the 79 health risk factors in 188 countries between 1990 and 2013 associates an average of about 5.5 million deaths per year due to the environmental contamination [5]. Finally, supporting also these facts, the European Agency for the Environment quantifies the premature deaths in the 28 European countries to 520 thousand during 2013 [6].

Given this apocalyptic scenario, it is key to reduce the emissions as much and as fast as possible. In this sense, the EU objectives for the reduction of emissions are quite ambitious compared to those of United States, China, or Japan. This is because it has pursued to reduce the CO_2 emissions from the current 130 g CO_2/km to just 95 g CO_2/km in 2020 and, furthermore, to 68–78 g CO_2/km in 2025 as shown in Fig. 1.

For this reason, many governments are supporting investments on EV charging infrastructure by means of incentives and subsidies as shown in Fig. 2 with the main goal of decarbonizing the transportation sector as soon as possible. Just as an example, Estonia has installed a public network of EV fast chargers along the whole country with more than 165 units with a maximum distance of 50 km between them and

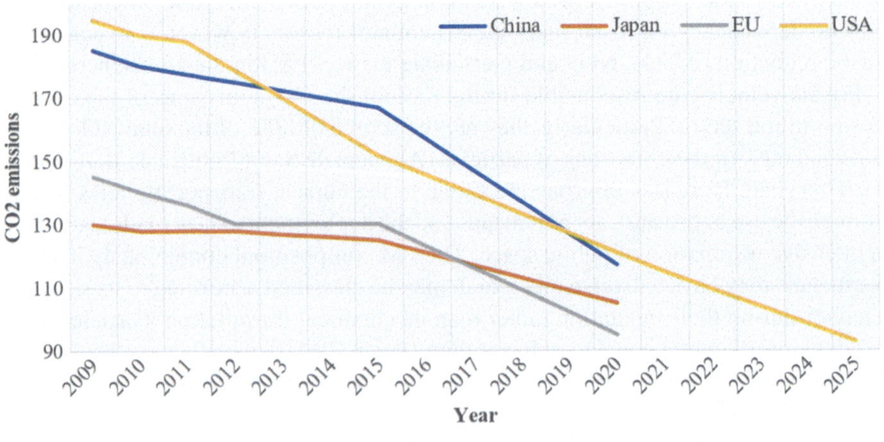

Fig. 1 Objectives for the reduction of CO_2 emissions in China, Japan, European Union, and United States. *Source* Mckinsey & Company [7]

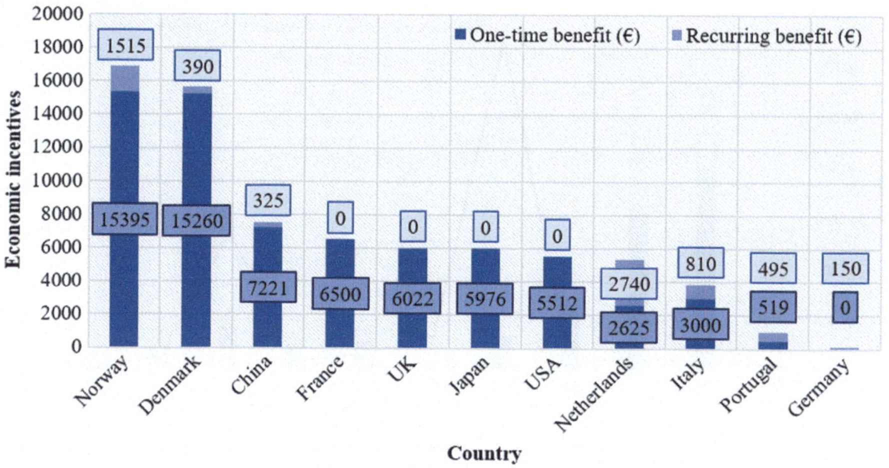

Fig. 2 National incentives to the EV purchase. Gray color: incentives to the investment. Blue color: incentives during a period after the purchase (tax reduction or other benefits). *Source* Adaptation from Mckinsey & Company [7]

having at least one EV fast charger in all the urban areas with more than 5 thousand inhabitants. Other countries like Norway, Denmark, the Netherlands, France, United Kingdom, and Spain are incentivizing their citizens to change to EVs using subsidies or incentives for their purchase.

Therefore, the actions for promoting a change from the traditional mobility concept is mainly based on internal combustion engines (ICE) to a sustainable one without a dependency on fossil fuels are diverse. For this reason, this section tries to analyze different international experiences which have been carried out with the main objective of fostering the e-mobility. The objective is to evaluate possible correlations in different countries between the type and quantity of the incentive or subsidy, shown in Fig. 2, with the EV penetration, as shown in Fig. 3.

2 EU Regulatory Context

Generally speaking, the European policy about the transportation sector embraces both regulatory issues (mandatory rules of procedure) and other documents regarding technology outlooks or feasibility reports. From 1996, but roundly from 2008, a wide set of actions have been taken to improve the quality of the air in the European Union. The following list enumerates those regulatory documents affecting the transportation sector including the Directive 2009/28/CE, which established the contribution of renewable energies to the gasoline, diesel, biofuels, and electric energy used in the transportation as well as the Rules of Procedure 333/2014 which limited the maximum emissions of the vehicles to 95 g CO_2/km in 2020 [10]:

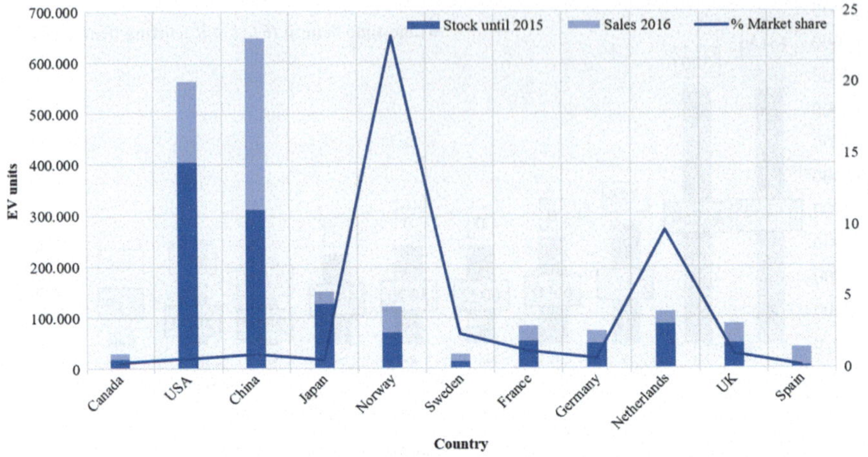

Fig. 3 EV market in different countries. *Source* OECD/IEA [8, 9]

- Regulation (EC) No 715/2007 of the European Parliament and of the Council of 20 June 2007 on type-approval of motor vehicles with respect to emissions from light passenger and commercial vehicles (Euro 5 and Euro 6) and on access to vehicle repair and maintenance information. Objective: It establishes the emission limits for light vehicles and the corresponding effective dates.
- Directive 2008/50/EC of the European Parliament and of the Council of 21 May 2008 on ambient air quality and cleaner air for Europe. Objective: It defines and establishes the main air quality characteristics to avoid, prevent, and reduce the negative effects to the human health and the environment.
- Commission Regulation (EC) No 692/2008 of 18 July 2008 implementing and amending Regulation (EC) No 715/2007 of the European Parliament and of the Council on type-approval of motor vehicles with respect to emissions from light passenger and commercial vehicles (Euro 5 and Euro 6) and on access to vehicle repair and maintenance information. Objective: It modifies the emission limits established in Regulation CE 715/2007.
- Directive 2009/28/EC of the European Parliament and of the Council of 23 April 2009 on the promotion of the use of energy from renewable sources and amending and subsequently repealing Directives 2001/77/EC and 2003/30/EC. Objective: If defines the mandatory national objectives for the renewable energy contribution in the final energy consumption and in transport consumption.
- Directive 2009/30/EC of the European Parliament and of the Council of 23 April 2009 amending Directive 98/70/EC as regards the specification of petrol, diesel, and gas-oil and introducing a mechanism to monitor and reduce greenhouse gas emissions and amending Council Directive 1999/32/EC as regards the specification of fuel used by inland waterway vessels and repealing Directive 93/12/EEC. Objective: It determines the technical specifications for the fuels used in ICEs

considering the technical requirements of these motors. Additionally, it defines an objective on the reduction of greenhouse gases during their life cycle.

- Directive 2009/33/EC of the European Parliament and of the Council of 23 April 2009 on the promotion of clean and energy-efficient road transport vehicles. Objective: It promotes the consideration of the energetic and environmental aspects associated with the useful life of vehicles. In this way, the market of clean and energy-efficient vehicles should be fostered in order to increase the positive impact of the transportation sector on the environmental, climate, and energy policies of the European Union.
- Regulation (EC) No 443/2009 of the European Parliament and of the Council of 23 April 2009 setting emission performance standards for new passenger cars as part of the Community's integrated approach to reduce CO_2 emissions from light-duty vehicles. Objective: It limits the average CO_2 emissions to 130 g/km for new light vehicles.
- Regulation (EU) No 333/2014 of the European Parliament and of the Council of 11 March 2014 amending Regulation (EC) No 443/2009 to define the modalities for reaching the 2020 target to reduce CO_2 emissions from new passenger cars. Objective: It limits the average CO_2 emissions to 95 g/km for new light vehicles in the market after 2020.
- Regulation (EU) No 1315/2013 of the European Parliament and of the Council of 11 December 2013 on Union guidelines for the development of the trans-European transport network and repealing Decision No 661/2010/EU. Objective: It establishes a set of recommendations for the development of a trans-European transport network with a double-layer structure.
- Directive 2014/94/EU of the European Parliament and of the Council of 22 October 2014 on the deployment of alternative fuels infrastructure. Objective: It sets a common frame for the deployment of alternative fuels with the aim of minimizing the oil dependency and mitigates the environmental impact of the transportation sector.
- Directive (EU) 2015/1513 of the European Parliament and of the Council of 9 September 2015 amending Directive 98/70/EC relating to the quality of petrol and diesel fuels and amending Directive 2009/28/EC on the promotion of the use of energy from renewable sources. Objective: It establishes the maximum amount of biofuels produced from crops of agricultural land and also considers the computation of electric power consumed by the electric vehicle from renewable energy sources.
- Commission Regulation (EU) 2016/427 of 10 March 2016 amending Regulation (EC) No 692/2008 as regards emissions from light passenger and commercial vehicles (Euro 6). Objective: It sets the maximum emissions considering actual driving conditions.
- Energy Efficiency Directive Winter package 2016, article 33 for the regularization of the functions of electric utilities on the management of EV chargers. Objective: It sets the possibility of giving up the utilities the management of electric vehicle charging infrastructure in case if other agents are not interested in this business.

In addition, each country must evaluate the potential interest that the involved stakeholders may have in this issue at least every five years.

2.1 Overview of Regulation About EV Charging in Different Countries

This subsection evaluates, in a descriptive manner, the different methods and strategies for boosting the EV in some European countries. It is important to take in mind that the economic status, the development level, the political situation, and the social awareness are key factors, among others, which determines the success or failure of an incentive to obtain the desired objective in a given country.

2.1.1 Denmark

The Danish government has the objective of eliminating fossil fuels by 2050. For this reason, the adopted measures have been aiming at deploying EV chargers and promoting infrastructure for hydrogen vehicles. This strategy is based, therefore, on promoting higher efficiency and clean vehicles. On the other hand, and additionally to these actions, the fuels by 2020 must contain at least 10% of biofuels.

In the Danish system with respect to the e-mobility, the public EV charging stations are being deployed by independent stakeholders which are not necessarily linked to the electric utilities. The construction, property, and operation of these EV charging infrastructure is, therefore, a competitive market where any person, company, or institution may deploy one or more EV charger in a given city and even in the same street.

Energinet, Danish TSO, undertook a study which evaluated the implementation of a platform for data exchange in 2007 and was developed later in 2009. This new information hub resulted in a decrement of the DSO responsibilities but a major involvement of the energy providers. Using this new system, the users contact directly with the energy provider without the need of a third-party agent [11].

Considering the incentives to the vehicle user, Denmark offered the exemption of the registration and circulation taxes for the electric and hybrid vehicles till 2015. Without any doubt, the objective of these measures is to attenuate the high purchasing cost of these kind of vehicles and to foster the competition with conventional vehicles based on internal combustion engines (ICE) [12]. Denmark has restrained the incentives for the EV purchase, but new supporting mechanisms are being revised. According to the International Energy Agency (IEA), Denmark is the only country that has revised downward their initial objectives in e-mobility for 2016.

Figure 4 shows the deployment of conventional and fast chargers in Denmark to provide a qualitative idea of the EV penetration in the country.

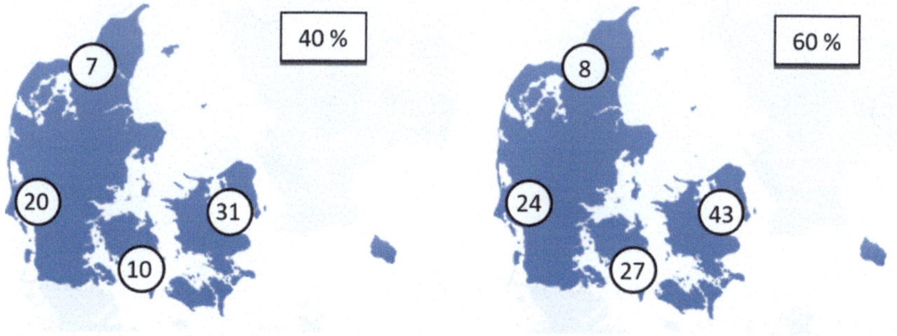

Fig. 4 Deployment of conventional and fast EV chargers in Denmark. Data from: Electromaps Borderless charging [13]

2.1.2 Estonia

The Estonian government commissioned to the finance and management agency KredEX to design a global nationwide e-mobility solution. Particularly, the location of EV chargers must be provided considering the required additional network reinforcements to stand the demand increase related to these new loads [12]. Estonia does not apply taxes to the purchase of vehicles. The government, however, considers an incentive for the EV purchases on the application of taxes to conventional ICE vehicles [14].

Estonia has a public EV charging infrastructure composed of 165 chargers in 2012. These EV chargers are based on the CHAdeMO standard and each village with more than 5.000 inhabitants has one at least. The maximum distance between EV chargers is between 40 and 60 km. In addition to this public charging infrastructure, the Estonian government offers economic support to individuals and companies in the EV purchase with amounts which may reach 18.000 €. Moreover, if an EV charger is installed at home this subsidy is incremented by 1.000 € [15].

This massive EV development and related charging infrastructure is the result of the agreement between Mitsubishi and the Estonian government signed in 2011. Mitsubishi acquired the emission rights of 10 million metric tons of equivalent CO_2 from Estonian emission allowances in 2011 in return of 507 i-MiEVs. In this way, Estonia promoted the use of EVs providing the EV fleets to public institutions, improving the charging infrastructure, and creating a program for incentives to the EV purchase [7].

The penetration of the EV is shown qualitatively in Fig. 5 through the deployment of conventional and fast EV chargers throughout the country.

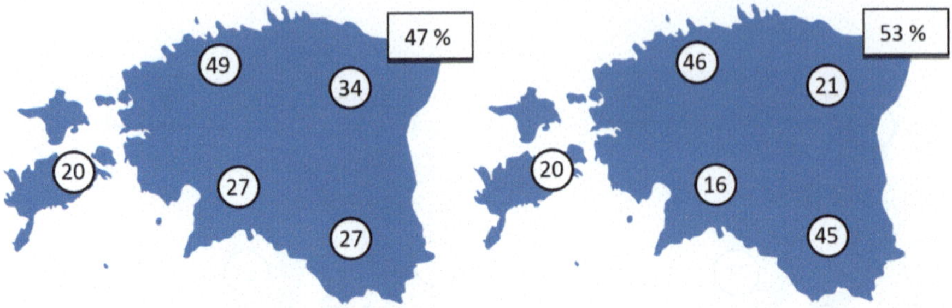

Fig. 5 Deployment of conventional and fast EV chargers in Estonia. Data from: Electromaps Borderless charging [13]

2.1.3 France

The French objective is to reach carbon neutrality by 2050 in addition to fulfill the European objectives of achieving a 40% reduction of emissions by 2030. In the short term, the objectives are to reduce the average emissions of new cars until 95 g CO_2/km for 2021, and to have a million of EVs and HEVs circulating as well as 1 charging point for each 10 ten EVs for 2022.

The incentives taken by the French government, among others, are for example the tax refund of 30% of the total cost of charging infrastructure in private housing, and the ADVENIR subsidies for collective housing, public area, company parkings, etc. In addition, some types of new buildings must have parking lots pre-equipped to facilitate the installation of charging points [16]. The deployment of EV chargers in France is shown in Fig. 6.

2.1.4 Germany

E-mobility is a clear national objective for Germany. The German development plan for e-mobility estates is a key objective to lead the EV deployment throughout the country but also to export EVs and its related technology. German manufacturers have invested and promoted on research and development but without forgetting critical issues regarding regulation and standardization. Probably this is the reason why Germany is a worldwide reference in this sector. Between 2009 and 2011, 500 million euros had been invested in e-mobility projects just to promote the development of the EV sector. In the next years, this policy continued with a total investment close to 1.000 million euros. The different promoting actions, among others, can be summarized as follows [12]:

- The establishment of a tax based on a fixed and variable quantities depending on the CO_2 emissions which are computed as a function of the engine capacity (2 €/100 cm^3 gasoline and 9.50 €/100 cm^3 diesel). The vehicles are exempt of taxes when the level of emissions is lower than a fixed limit, recently reduced from

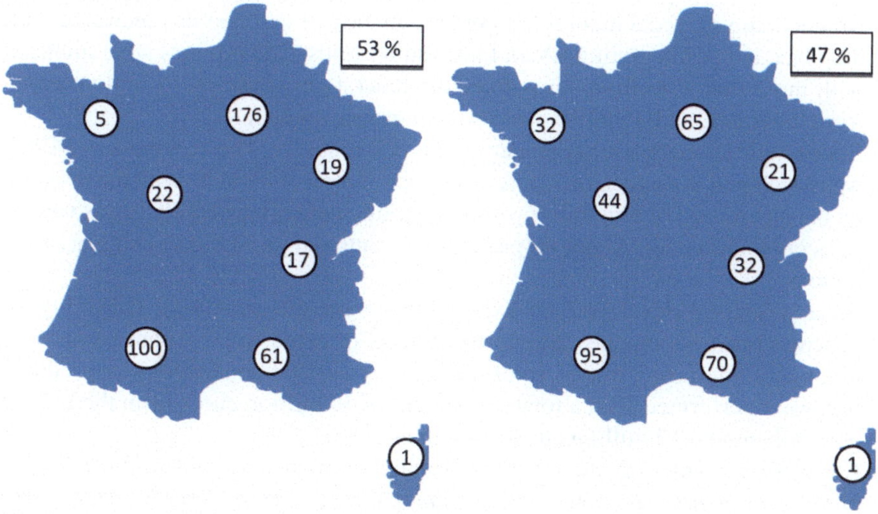

Fig. 6 Deployment of conventional and fast EV chargers in France. Data from: Electromaps Borderless charging [13]

100 to 95 g/km. This tax exemption applies during 10 years for vehicles acquired before 2015 but just 5 years for those ones acquired before 2020.

- Initially, it has been announced in 2010 that the EV final purchase price should not receive any reduction through subsidies. In June 2013, however, it has been approved as a law which offered economic incentives as a function of the battery size: 500 €/kWh with a maximum reduction of 10.000 € and decaying 50 €/kWh every year [17].

In addition, other promoting programs were running in parallel as the "Electromobility Model Regions" sponsored by the Federal Ministry of Transport and Digital Infrastructure where eight target regions were subsidized with 130 million euros in 2009. Different experts from scientific, industrial, and local governance fields cooperated to build and operate the required infrastructure for e-mobility.

Later in April 2012, German government selected four regions (Berlin-Brandenburg, Baden-Wurttemberg, Bavaria-Saxony, and Lower Saxony) as test regions for large-scale demonstrative projects and pilot projects with a total investment over 180 million euros. In addition to this national public investment, regional governments also participated supporting the initiative with more than 80 million euros. Actions in each of the regions can be summarized as follows [18]:

- Berlin-Brandenburg. The project was coordinated by the Berlin e-mobility Agency (eMO) which developed more than 30 projects with more than 100 partners (industry, university, municipalities, etc.). Their goals were focused on the driving experience, parking, charging, energy storage, and technology integration. With more than 1.500 eVs and 400 charging stations, Berlin was leading the EV development in Germany regarding the practical integration of e-mobility, while

Brandenburg played a main role in the production of sustainable renewable with the objective of integrating EV in future smart grids. The projects were financed with more than 7 million euros from the federal government, 19 million euros from the states, and 35 million euros from private companies.
- Baden-Wurttemberg was a pioneer region in e-mobility. The "Living Lab MWe" was developed in this region where a network of projects dealing with the research on the integration of e-mobility within the traffic network and daily life as well as its economic impact. About 40 projects concentrated on Stuttgart and Karlsruhe regions.
- Bavaria-Saxony developed projects focusing on the economic feasibility of different business models. About 40 projects oriented to large-range journeys, urban and interurban routes, rural traffic, international connections, and education were performed with a total investment of 39 million euros from the federal government and 15 million euros from the states.
- Lower Saxony (metropolitan region of Hannover Braunschweig Göttingen Wolfsburg) is a major area on vehicle manufacturing. Thirty-four projects were developed focusing on interoperability, new concepts of charging infrastructure, renewable energies, EV, and related components manufacturing.

All these actions have achieved a large EV deployment which can be shown in Fig. 7 where the conventional and fast chargers along Germany are represented.

2.1.5 Netherlands

The Netherlands pursues the objective that by 2020 at least 10% of the purchased vehicles were based on either hybrid or pure electric technologies. This objective is even more ambitious for 2025 because this target is incremented to 50% with at least 30% of pure electric vehicles. For doing so, it is crucial to count with an outstanding national network of EV charging infrastructure.

Subsidies, in addition, to provide some financial aid to the EV owners, are also provided to a greater extent for deploying the EV charging infrastructure. Netherlands also stimulated the EV market development, the innovation through living labs without compromising tax incentives and a continuous supervision of the sector evolution to take corrective actions if needed. The tax incentives approved by 2015 are summarized as follows [19, 20]:

- Purchase car tax depending on the emissions:
  - Zero-emission vehicle without taxes.
  - 2 €/g for vehicles with emissions between 1 and 76 g CO_2/km.
  - 66 €/g for vehicles with emissions between 77 and 102 g CO_2/km.
  - 145 €/g for vehicles with emissions between 103 and 150 g CO_2/km.
  - 238 €/g for vehicles with emissions between 151 and 168 g CO_2/km.
  - 475 €/g for vehicles with emissions above 169 g CO_2/km.

- Exemption of circulation tax:
  - Without taxes for vehicles below 50 g CO_2/km.
  - Between 400 and 1.200 € for higher emissions and depending on the type of fuel, vehicles weight, and mass.
- Reduction of the tax related to the private use of a company vehicle: 4% for a pure electric EV and 7% for PHEV instead of the 14–25% applied for ICE vehicles.
- Up to 4.500 € of reimbursement on the purchase price due to the positive environmental impact.
- Other local incentives.

Due to these actions, Table 1, Figs. 8 and 9 show the increment of EV vehicles, EV charging stations, and their distribution within the Netherlands, respectively.

2.1.6 Italy

Italy started the EV integration in 2010 when AEEGSI (Authority for the electric energy, gas and water system) decided to evaluate different business models for the EV charging in public places [21]. For this reason, it has been promoted some

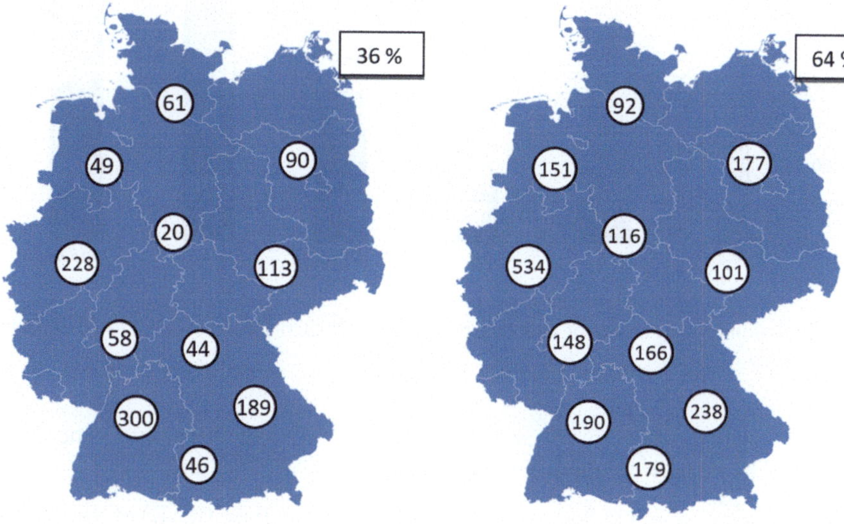

Fig. 7 Deployment of conventional and fast EV chargers in Germany. Data from: Electromaps Borderless charging [13]

Table 1 Increment of EVs as a function of the available charging infrastructure installed for three years

Vehicle	31-12-2013	31-12-2014	30-09-2015
Passenger car (EV)	4161	6825	9038
Passenger car (E-REV, PHEV)[a]	24512	36937	53165
Commercial car <3.5 t	669	1258	1544
Commercial car >3.5 t	39	46	50
Bus[b]	73	80	95
Tricycle	632	769	847
Motorbike	125	196	278
Total	30211	46111	64928

[a]Excluding full-hybrid vehicles; [b]Including trolley busses and some hybrid busses
Adaptation from Nieuwenhuis [20]

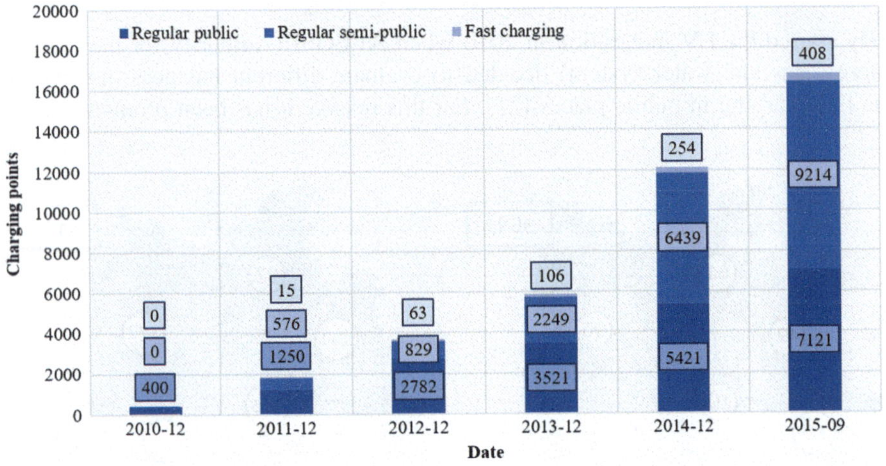

Fig. 8 Evolution of the number of charging points in the Netherlands from 2010 to 2015. Data extracted from [20]

project calls with the objective of evaluating possible different alternatives. The main problems to be faced were the following:

- EV chargers were not included in the Regulatory Asset Base (RAB) of the DSOs. Therefore, it was proposed to remunerate their installation 728 € per EV charger and year.
- In addition, conventional slow charge was the unique mature existing technology at that moment.
- The domestic EV charge was a problem because the existing electricity tariffs were introduced in the seventies of the last century is based on a progressive scheme (energy term rises with the energy consumption) with the objective of

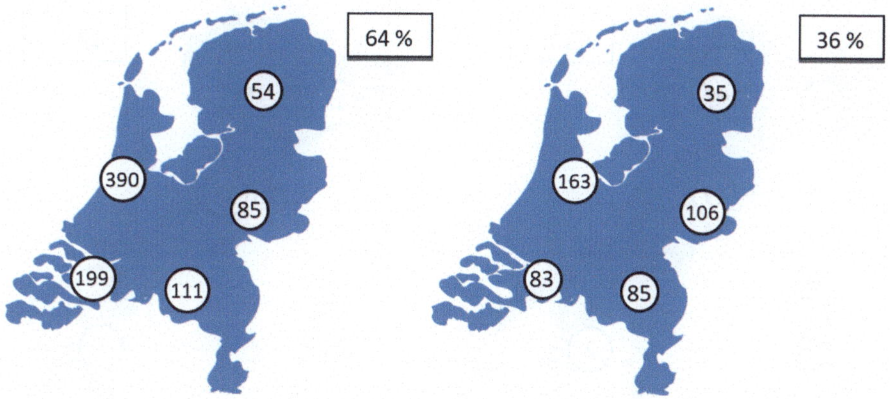

Fig. 9 Deployment of conventional and fast EV chargers in the Netherlands. Data from: Electromaps Borderless charging [13]

incentivizing the efficiency energy. For this reason, AEEGSI approved a change in the electricity tariff in such a way that the final users may use the domestic EV charge with a competitive cost.

Considering these constraints, only 5 projects were selected (one or two per each business model). These pilot projects have the initial objective of installing up to 500 EV conventional chargers in public places of nine Italian regions but following different business models which will be analyzed later on Sect. 2.3:

- DSO owner of the technical and commercial management of the EV chargers. This case was tested in Pisa, Bari, Geneve, Perugia, Emilia-Romagna, and Milan with 310 eV chargers.
- Area-licensed model was tested in Milan and Brescia with 100 eV chargers.
- Competitive energy providers model was tested in Milan and Rome by Enel with 26 eV chargers but also Clas Onlus installed 150 eV chargers in Roma, Milan, Bari, Catania, Geneve, Bologna, and Varese.

A view of the charging stations distribution along Italy is depicted in Fig. 10.

2.1.7 Norway

Norway is one of the most advanced countries in e-mobility and for this reason one of the models to pursue. To increase the number of EVs is defined as a key action of the Norwegian Climatic Policy and, therefore, Norway has been used a wide range of incentives to turn EVs as a competitive and attractive option. The actions applied to pure electric EV are summarized as follows [22]:

- VAT exemption in the EV purchase (2001). Due to the fact of their higher production costs, the VAT related to EV is really high yielding to an excessive price

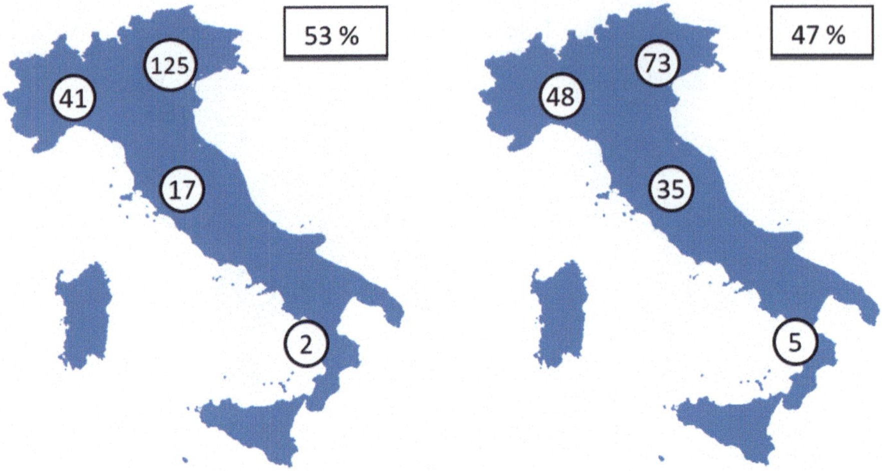

Fig. 10 Deployment of conventional and fast EV chargers in Italy. Data from: Electromaps Borderless charging [13]

difference with respect to traditional ICE vehicles. In spite of this tax exemption reduces the tax revenue, it has a positive effect on the EV sales.

- Access to bus lanes (2003/2005). This action turns the EV more convenient than conventional ICE ones in cities with traffic jams. However, this incentive can be a little bit risky because an increment on the number of EVs circulating within the bus lane may create delays in the busses.
- Registration tax exemption (1900/1996). In the beginning, these actions were just applied for EVs with a purchase price above a given value but, after that, it has been applied to every EV. Note that the circulation tax in Norway is between 2.600 and 9.400 € depending on the vehicle characteristics.
- Free parking. This action is especially effective in zones with limited parking areas. The influence on the number of EVs is even more visible with the reservation of parking zones exclusively for EVs.
- Free toll (1997). In the Oslo area, the tolls may reach about 600–1.000 € every year. However, this amount increases to 2.500 € per year in remote areas due to the tunnel maintenance. This incentive has motivated an increment of the EV fleet in island areas.
- Circulation tax reduction (1996–2004). EVs payed a minimum of about 52 € while ICE vehicles payed between 360 and 420 €.
- Reduction in the price of ferries (2009). The success of this action had been really limited.
- Reduction on the corporate taxes of company vehicles (2000). It has almost no impact till 2012 but it started being interesting with the Tesla Model S launching.
- Financial support to the EV charging stations (2009). In this way, it reduced the financial risk and, therefore, it has supported the installation of EV charging

Table 2 Yearly subsidies for the EV customer. Norwegian fleet: 25,000 BEVs in April 2014

Incentive	Amount per vehicle (€/year) (2014)
Free tolls	434
Bus lane	940
Free parking	398
Free ferries	145
Total	1.928

Source Assum et al. [22]

stations. This has motivated a reduction of the driver concerns about the EV autonomy which has motivated an increment of EV sales.

- EV fast charging stations (2011). Fast charging increases the EV miles driven and the total EV market. It becomes easier for fleets to use EVs and is a premise for using EVs as taxis.
- Reserved car license numbers (1999). This facilitates the control of other incentives such as free parking and toll.

Note that all these actions are for pure EVs but in the case of hybrid ones just two of them applies: reduction of the registration tax and free charge on public charging places. Without any doubt, the aggregated effect of all of these actions implies great economic benefits for the EV users as shown in Table 2 [22].

Complementarily, the project COMPETT had also computed the average economic value of the local incentives for the pure EV which is about 1.928 per vehicle (a total of 48.5 million euros). Finally, it is important to remark the key role that driver associations, like Norsk Elbilforening, had played in the EV deployment in Norway. With all of these actions the deployment of EV charging infrastructure in Norway is shown in Fig. 11.

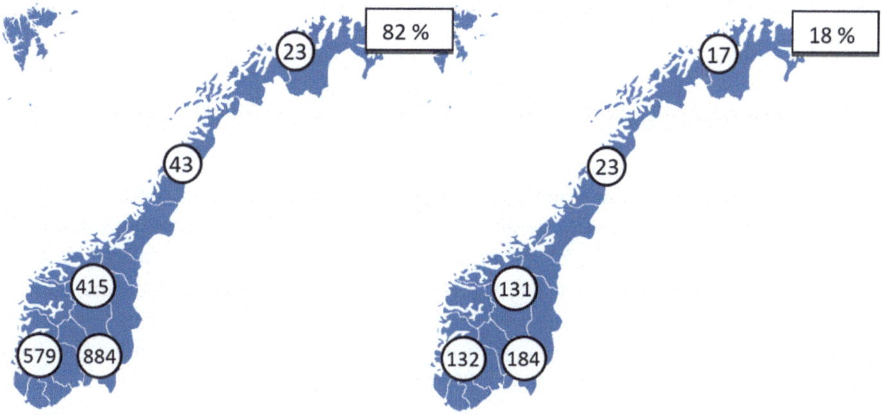

Fig. 11 Deployment of conventional and fast EV chargers in Norway. Data from: Electromaps Borderless charging [13]

Table 3 Summary of incentives taken by different countries

Country	EV purchase support	Charging points infrastructure support	IC vehicles taxes	EV taxes reduction	Manufacturers support
Denmark	✓	✓		✓	
Estonia	✓	✓	✓		
France	✓	✓		✓	
Italy	✓			✓	
Germany	✓	✓		✓	✓
Norway	✓	✓		✓	
Netherlands	✓	✓	✓	✓	

Table 4 Density of all chargers and fast types that are operative nowadays. Data extracted from OECD/IEA [8]

Country	Chargers per km^2	Fast chargers per km^2
Denmark	0.0030	0.0019
Estonia	0.0065	0.0030
France	0.0016	0.00034
Germany	0.0088	0.0047
Netherlands	0.0322	0.0117
Italy	0.0018	0.00051
Norway	0.0096	0.0018
Spain	0.0047	0.0011

Table 3 summarizes the different incentives and actions analyzed in the previous sections. Finally, and in order to evaluate their impact, Table 4 shows the EV density (EV chargers/km^2) in the analyzed countries.

2.2 *E-Mobility Stakeholders*

E-mobility market gets the attention of multiple stakeholders. Some of them are directly related, such as the EV owners, EV and charge infrastructure suppliers, EV charging unit operators, and e-mobility service providers. Other stakeholders belong to sectors that, if not directly related, have a close relationship such as the energy providers, Distribution System Operators (DSOs), Transmissions System operators (TSOs), and clearing houses. A description and the relationship between the involved stakeholders are summarized in Fig. 12 [23].

EV users are the most important clients of e-mobility, and with the objective of achieving a EV demand as high as possible, the actions must be oriented to satisfy all

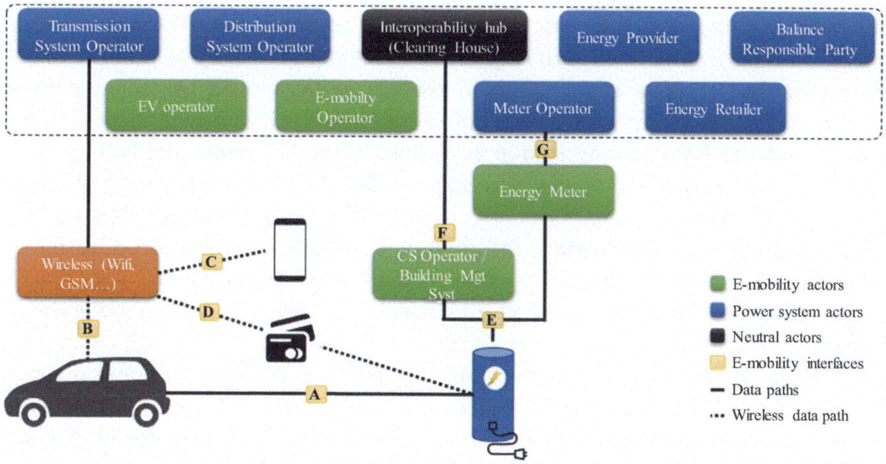

Fig. 12 Scheme of the different agents taking part in the EV charging market. *Source* Zabala et al. [23]

their needs: (i) productive, comfortable, and efficient use through a remote communication system between the charging infrastructure and the vehicle; (ii) battery charge before any travel enough to reach a give destination; (iii) interoperability between, at least, European countries.

EV and equipment manufacturers are devoted to bring the vehicles and all the required auxiliary equipment (EV chargers, connectors, etc.) to the market. They must fulfill a set of standards and regulatory issues with the objective of achieving the interoperability between the European countries. In this way, EV owners may travel pleasantly with the absolute certainty that they may use the charging infrastructure of any country without problems. However, and complying with all the standards and regulatory issues, manufacturers are able to offer different technological solutions to a common problem. Therefore, sometimes interoperability is not always fully guaranteed. For this reason, it is key to have information about the EV users and other stakeholders to raise technological solutions aligned with the requirements and expectations of all the involved agents.

EV charging station operators are responsible for operating and maintaining this asset. They just operate EV charging stations meaning that they do not have any continuous contractual relationship with the EV owners. They are in charge of buying the energy provided to the EVs but also they must manage the user identification, authorization, and payments of EV charging as well as guaranteeing the quality of the provided services. For this reason, they must adapt to the emerging technologies, tools for measuring and monitoring some electric network parameters and communication infrastructure to share information with other stakeholders as DSO, energy provider, e-mobility service provider, and EVs.

The e-mobility service providers are legal entities which maintain a contract with an EV user for all the services related to the EV charging. They are stakeholders with the capability of identifying users with their personal data. Their main functions are wide, ranging from verifying the contract validity, charging authorization, data sharing between the stakeholders to provide services to the users and support to the logistic management, providing information to EV owner with respect to the most convenient EV charging station according to the EV range, the battery state of charge, the availability of the charging station, the energy price, the distribution network congestion, etc. Evidently, these e-mobility service providers require communication with third parties, information about the identification of different stakeholders on a given charging operation, and location information. Note that to maintain interoperability is a key factor as e-mobility service providers are in direct contact with the final EV users.

DSOs are the entities responsible for the operation, development, and maintenance of the distribution network. In the future, it is expected that DSOs manage the integration of distributed generation and control loads in an active way. Among the DSO responsibilities with respect to the e-mobility, the management of their assets to reduce as much as possible network congestions, a secure energy supply, restoration capability after a supply outage, bidirectional communication with other stakeholders, management of some ancillary services required for real-time control of the network, and investments on new network assets stand out. In addition, DSO may facilitate the integration of smart devices to assure the system stability. In this regard, considering that nowadays most of the distribution systems are passive ones, the deployment of new smart grid technologies like Demand Side Management (DSM), the Vehicle to Grid (V2G) operation, or the joint integration of renewable energies and EV are key issues for guaranteeing the stability, controllability, and reliability of the distribution network.

Finally, the energy provider is the entity which sells the energy to the final user according to the regulatory framework. EV charging can be done by purchasing the energy to and energy provider or to a e-mobility service provider. The energy provider requires to manage the EV charging operation, communication interfaces with the e-mobility providers, and EV charging operations as well as the EV user.

2.3 DSO Roles

At the beginning of the EV deployment in Italy, it has been proposed three types of business models with different approaches in the way that distribution, charging, and energy supply were carried out by separated stakeholders as shown in Table 5.

In the first model, the DSO assumes the technical management of the EV charging stations with different energy suppliers which can be selected by the final users depending on their preferences. However, this model was considered incompatible after the Directive 94/2014/UE on the deployment of alternative fuels infrastructure [21]. The second option creates an intermediate agent between the utility and the energy supplier which manages technically the charging stations. Finally, the third option joints the techno-economic management in the figure of an integrated charging

Table 5 Different business models initially evaluated by the Italian government in 2010, in which the distribution, technical operation, and commercial operation lay on different agents. Fuente: Schiavo [21]

Business models	Distribution	Recharging	Supply
DSO	Distribution company *with accounting separation*		Retail suppliers
Area-licensed Service provider	Distribution company	Rech. Provider (*local license*)	Retail suppliers
Competitive service provider	Distribution company	Integrated recharge providers (*in competition like fuel stations*)	

agent with a similar role as of the current petrol stations. The main problem is that the high infrastructure cost and the actual reduced number of EVs result in a low business profitability. However, it can be found alternative models promoted by private companies as Tesla o public municipalities which installed EV charging points as a public service.

Currently countries like Austria, Luxemburg, Slovenia, and Ireland have adopted a model where DSOs are the owners of the EV charging infrastructure being responsible for their operation following a natural extension of their traditional role. However, punctual differences arise in some implementation issues. In Austria, Luxemburg, and Slovenia the DSO is responsible for the metering infrastructure, technical operation, and commercial management. Ireland, on the contrary, segregates the metering billing and commercial management which depend on a third-party agent but the DSO owns the EV charging stations which are an additional asset in the distribution network. This model is fully compatible with the European Directive 2014/94/EC, textually: "Distribution system operators play an important role in relation to recharging points. In the development of their tasks, the distribution system operators, some of whom may be part of a vertically integrated undertaking owning or operating recharging points, should cooperate on a non-discriminatory basis with any other owners or operators of recharging points, in particular providing them with the information needed for the efficient access to and use of the system." In addition, it also follows the guideline of the document [24] emitted by Council of European Energy Regulators (CEER), textually: "When there is the potential for competition to develop new activity areas, the default is either to prevent DSOs from undertaking the activity completely, or allow the DSO to undertake the activity under special conditions imposed by the regulator." Finally, the 2016 Winter Package in article 33 establishes that the state members may give the ownership, development, and management of the EV infrastructure just in case other stakeholders were not interested in it. Moreover, every five years the state members must re-evaluate the potential interest that the other stakeholders may have on this issue. Later on, the European Federation of Local Energy Companies (CEDEC) proposed an amendment emphasizing that the management of the EV charging infrastructure must offer a competitive price and also cover all the geographic areas where the DSO operates.

In the opinion of Eurelectric, as the market grows toward a more competitive one the DSOs may release this responsibility. Meanwhile, DSOs may be the owners and technical managers of the EV charging infrastructure, but it will be required to elaborate a strategy to allow the incorporation of new stakeholders when the market maturity arises. During this initial period, the investment return could be done by either adding in the RAB these new assets or by public financial resources. This last option should allow an equitable share of decarbonization cost in the society. Once the market maturity happens it should be possible to incorporate new competitive stakeholders to avoid DSO stranded costs as shown in Fig. 13. In any case, the regulatory authority should perform a market analysis to determine its convenience. DSO may maintain the ownership of the EV infrastructure up to the complete return on the investment or may have the opportunity to sell it at its residual cost [25].

Nevertheless, alternatives schemes are also possible where the DSO functions are lower such in the case of some Nordic countries and France. In these countries, the public EV infrastructure deployment is taken by other stakeholders. The new EV stations are considered as a conventional network extension. In this model, namely independent e-mobility, the DSO has the role of an information hub. Its main function is to gather and distribute the information among the involved stakeholders [26].

Table 6 summarizes the different roles that DSO may have with respect to the EV charging infrastructure in different European countries.

Fig. 13 Evolution of charging business according to Eurelectric. *Source* Eurelectric [25]

Table 6 Different roles that the DSO may have in different European countries in regards to the EV infrastructure

Country	Commercial operation	Charging points investment and technical operation	Information Hub
Austria	✓	✓	–
Luxembourg	✓	✓	–
Slovenia (highways)	✓	✓	–
Ireland	–	✓	–
Denmark	–	–	✓
Norway	–	–	✓
France	–	–	✓
Spain	–	–	✓
Italy	✓	✓	–
Germany	–	–	✓

2.4 *Interoperability and Business Models*

The regulatory framework has a direct impact on the activity within a region with regard to the EV deployment. It is mandatory, therefore, that the regulation promotes profitable and sustainable business models. Issues like the competitiveness protection, taxes, and administrative barriers, support and benefit the final user to have a direct relationship with the EV penetration. Other issues like the operational procedures of distribution network, access tariffs, market requirements, safety, and environmental rules, etc. do not have such a direct impact but also influence the business.

In any case, the business models related to the interoperability are based on providing services. Most of them have been tested in pilot projects but are still required to bring them to the market. The EV charging service could be offered to the final user in different ways [23]:

- Free access: EV users do not need a previous contract with the Enterprise Mobility Service Provider (EMSP). It is the EV charging station operator who has contracts with some EMSPs and offers the user to choose one among them.
- Without roaming: EV users only may recharge in those EV charging stations with an agreement with their own EMSP.
- With roaming: EV users may charge in EV charging stations without a direct relationship with their own EMSP. In this case, it is required the participation of a Clearing House.
- Private charging.

Each of these options has advantages and disadvantages. For instance, free access option has low information requirements because of the reduced number of the involved stakeholders. In the roaming option, it is required the participation of additional agents like the Clearing House, but the process is quite simple because the information is centralized. In any case, this is a competitive service market, therefore, companies will offer new services to provide added value with respect to their competence.

References

1. Agència de Salut Pública de Barcelona (2012) Los contaminantes atmosféricos y la salud: las partículas en suspensión (PM)
2. Drisse M-NB, Goldizen F, Adair-Rohani H (2017) Atlas on children's health and the environment
3. World Health Organization (2017) Don't Pollute my future! the impact of the environment on children's health. World Health Organization, vol 30. https://doi.org/10.1021/bi00088a015

4. World Health Organization (2014) Burden of disease from household air pollution for 2012. Summary of results. In: World Health Organization, vol 35. http://www.who.int/phe/health_topics/outdoorair/databases/FINAL_HAP_AAP_BoD_24March2014.pdf
5. Forouzanfar MH, Alexander L, Anderson HR, Bachman VF, Biryukov S, Brauer M et al (2015) Global, regional, and national comparative risk assessment of 79 behavioural, environmental and occupational, and metabolic risks or clusters of risks in 188 countries, 1990–2013: a systematic analysis for the Global Burden of Disease Study 2013. Lancet (London, England) 386(10010):2287–2323. https://doi.org/10.1016/S0140-6736(15)00128-2
6. Planelles M (2016) 30.000 muertes prematuras al año en España por la contaminación. El País. https://elpais.com/politica/2016/11/23/actualidad/1479905871_541192.html
7. Mckinsey & Company (2012) Electric vehicles in Europe. https://www.google.co.uk/url?sa=t&rct=j&q=&esrc=s&source=web&cd=2&cad=rja&uact=8&ved=0CC0QFjAB&url, http://www.mckinsey.com/~/media/McKinseyOffices/Netherlands/Latestthinking/PDFs/Electric-Vehicle-Report-EN_ASFINAL.ashx&ei=7MUWVf2zNcTeU42lg-AC&usg
8. OECD/IEA (2016) Global EV outlook 2016—Beyond one million electric cars, paris: international Energy Agency
9. OECD/IEA (2017) Global EV outlook 2017. https://www.iea.org/reports/global-ev-outlook-2017
10. Álvarez Pelegry E, Menéndez Sánchez J (2017) Energías alternativas para el transporte de pasajeros - Orkestra Instituto Vasco de Competitividad. https://www.orkestra.deusto.es/es/investigacion/publicaciones/cuadernos-orkestra/1150-energias-alternativas-transporte-pasajeros
11. THEMA Consulting group (2015) Mapping of TSO and DSO roles and responsibilities related to information exchange. http://www.nordicenergyregulators.org/wp-content/uploads/2015/06/THEMA-Report-2015-02-Mapping-of-TSO-and-DSO-responsibilities-related-to-information-exchange.pdf
12. BASREC (2015) The development of electric transport—Its effect on the security of the electrical energy system and forecasting energy demand in chosen 8 BASREC countries (Norway, Denmark, Germany, Sweden, Finland, Estonia, Lithuania and Poland). http://basrec.net/wp-content/uploads/2016/01/Emobility_10_11_2015.pdf
13. Electromaps Borderless charging (2017) Electromaps. www.electromaps.com
14. Sahuquillo M (2017) How Estonia persuaded drivers to go electric. El País. https://english.elpais.com/elpais/2017/05/29/inenglish/1496070465_626777.html
15. Estonian World (2013) Estonia becomes the first in the world to open a nationwide electric vehicle fast-charging network. Estonian world. https://estonianworld.com/technology/estonia-becomes-the-first-in-the-world-to-open-a-nationwide-electric-vehicle-fast-charging-network/
16. Lagarde D (2019) Public infrastructure for electric mobility in France
17. Reuters (2013) Germany ends tax disadvantage for corporate electric cars. Reuters. http://www.reuters.com/article/germany-electricvehicles-tax-idUSL5N0EJ1Y320130607
18. VDI/VDE IT & TÜV Rheinland (2017) Projektträger Schaufenster Elektromobilität. Projektträger Schaufenster Elektromobilität. http://schaufenster-elektromobilitaet.org/en/content/ueber_das_programm/foerderung_schaufensterprogramm/foerderung_schaufensterprogramm_1.html
19. Jonnaert E (2017) 2017 ACEA tax guide. Acea 1(1):372. http://www.acea.be/uploads/news_documents/ACEA_Tax_Guide_2017.pdf
20. Nieuwenhuis J (2015) E-mobility in the Netherlands, p 5
21. Schiavo Lo L (2017) Competition and regulatory aspects of electric vehicles charging: the Italian regulatory experience, vol 10
22. Assum T, Kolbenstvedt M, Figenbaum E (2014) The future of electromobility in Norway—Some stakeholder perspectives. In: TØI report. ISBN: 978-82-480-1602-1
23. Zabala E, Rodriguez R, Ratej J, Misara S, Kelm P, Olek B et al (2017) Business opportunities for interoperability assessment of integration

24. CEER (2016) 2016 Annual Report of the council of European Energy Regulators (CEER)
25. Eurelectric (2016) Charging infrastructure for electric vehicles. In: Eurelectric, Electricity for Europe
26. Ecorys (2014) The role of DSOs in a smart grid environment. Final Report Project Ecorys and ECN for DG ENER. https://www.ecn.nl/publications/PdfFetch.aspx?nr=ECN-O-14-031

Forecast of EV Derived Electrical Demand. The Spanish Case

Francisco José Gutiérrez-García and Ángel Arcos-Vargas

Abstract The inclusion of the electric vehicles in Spain is unavoidable, as well as in the developed countries, and the electrification of the Spanish vehicle fleet will produce an increase in the electric energy demand which will require changes in the current infrastructure. This chapter estimates the total consumption of an entire electric fleet considering the electrification of all the vehicles registered in Spain. The analysis performs a sampling of the most recent models of EV in the market and the typical driven annual distances of each type of vehicle. Finally, the environmental impact is estimated regarding CO_2 emissions, the optimal charging scheme is established and some consideration about the impact in the grid are taken into account.

1 Introduction

This chapter is focused on carrying out an exhaustive analysis of the EV scenario in Spain. An assessment of the current ground transport by road is performed in order to establish a comparison with the ideal case of a full electrification of the Spanish vehicle fleet.

First of all, the vehicle fleet of Spain is analysed for every city and town of the country with the aim to establish the number of vehicles of each type (car, truck, busses, motorbikes, etc.). Subsequently, it is necessary to estimate the consumption of each EV type, both specific electric consumption and the average travelled distance, to calculate the total electric demand. Once the total electric consumption is calculated, the hourly demand curves are applied to characterize the consumption as a function of time which allows to estimate the impact in the current energy infrastructure and define typical charging schemes and scenarios.

Finally, the environmental impact is estimated related to CO_2 emissions, comparing the consumption of the current vehicles with the new electric demand which has to be supplied with electric energy generation systems.

F. J. Gutiérrez-García · Á. Arcos-Vargas (✉)
University of Seville, Seville, Spain
e-mail: aarcos@us.es

A. Arcos-Vargas (ed.), *The Role of the Electric Vehicle in the Energy Transition*,
Green Energy and Technology, https://doi.org/10.1007/978-3-030-50633-9_2

2 Data

Ground vehicles can be divided into two mains categories: vehicles whose direction is restricted by a track set, for example, the railway vehicles, and vehicles without any restriction in their movement, free to move in any direction. This assessment only considers the second category, free-to-move vehicles, since vehicles with restricted direction (train, tram, etc) can be electrified with an own fixed infrastructure deployed along their route, which provides them a continuous electric energy supply.

The second category embraces a big variety of vehicles, from the simple case of non-motorized vehicles like a bike to a heavy machine like a cement mixer. However, when talking about electrification of the vehicles, only conventional vehicles are considered to be displaced in favor of electric vehicles, excluding non-motorized vehicles and heavy machines. The reason of that is the uncertainty of how many and which type of vehicles will be electrified in the next decades. So, the vehicles considered for the electrification are the following: cars, vans, buses, trucks and motorcycles.

2.1 Spanish Vehicle Fleet

The current Spanish vehicle fleet has a total of 33.7 million of vehicles which are registered and paying taxes according to the Dirección General de Tráfico [1]. This number of vehicles included all of them without considering the type.

Among the 33.7 million of vehicles, 32.6 million are included in the types considered before for electrification and 29.8 million belong to the municipalities of the continental Spain (Without Canary and Balearic Islands nor North-African cities). This exclusion is made under the objective of analysing the impact of the EV inclusion into the real infrastructure of Spain, avoiding the specific ones of the excluded regions mentioned above.

Table 1 summarizes the number of vehicles by categories for the 15 continental regions of Spain.

As can be seen in Table 1, cars lead the Spanish vehicle fleet with over 22 million followed by vans, motorcycles and trucks with about the same quantity (2–3 million) for each one. Finally, buses comprise only a small part of the fleet with a total of 56,071 units.

An interesting issue to point out is the age of the vehicles in order to show the obsolete technology deployed in the Spanish fleet. The older car registered dates from 1901 and the older truck from 1902. Moreover, there are about 43,000 vehicles registered before 1960. The number of vehicles on the roads and registered since every year from 1960 to 2018 is shown in Fig. 1.

Taking into account the age of the Spanish fleet vehicles shown in the figure (About 79% of vehicles registered since 2010) and the inclusion of the EV in Spain, the electrification of the common transport is supposed to be for the entire fleet.

Table 1 Spanish fleet

CCAA	Cars	Vans	Buses	Trucks	Motorcycles	Total
Andalucía	4,080,704	412,476	9,294	462,897	655,261	5,620,632
Aragón	608,550	70,529	1,510	73,541	76,624	830,754
Asturias	516,400	46,942	1,455	40,402	57,057	662,256
Cantabria	305,540	25,170	633	31,120	39,225	401,688
Castilla La Mancha	1,073,946	129,610	2,318	150,543	112,011	1,468,428
Castilla y León	1,308,808	135,096	3,286	138,961	131,829	1,717,980
Cataluña	3,527,529	387,006	9,361	374,515	834,704	5,133,115
Comunidad Valenciana	2,567,237	198,275	4,599	277,858	386,780	3,434,749
Extremadura	589,861	71,163	1,364	68,125	55,018	785,531
Galicia	1,538,995	116,219	4,742	134,881	158,605	1,953,442
La Rioja	147,630	17,458	263	20,603	16,633	202,587
Madrid	3,759,902	394,281	11,040	260,658	364,437	4,790,318
Navarra	326,338	39,010	849	37,799	35,386	439,382
País Vasco	1,002,752	83,124	3,475	94,142	131,951	1,315,444
Región de Murcia	759,531	66,871	1,882	86,380	111,279	1,025,943
Total	22,113,723	2,193,230	56,071	2,252,425	3,166,800	29,782,249

Source DGT [1]

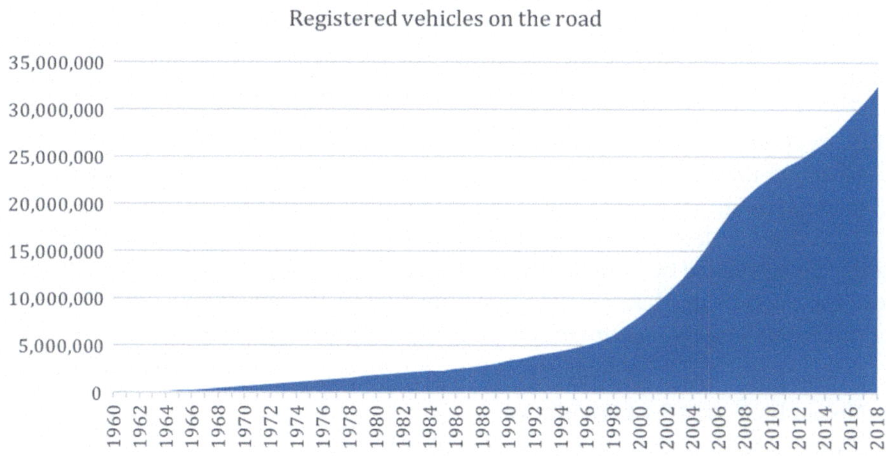

Fig. 1 Registered vehicles on road. *Source* DGT [1]

2.2 *Use and Consumption*

Like the fuel consumption (l/100 km) is for the standard combustion vehicles, the electric energy consumption is for EV. This electric consumption is measured in kWh/100 km and it is an indicator which represents the energy necessary to achieve a certain range of distance. This way, this specification allows to compare different type of vehicles regardless other parameters as weight, sizes, capacity, etc., being a common indicator for every EV.

The consumption for each type of vehicle (cars, vans, buses, etc.) is estimated by a sample of several models in the current market for each case in order to obtain a value which actually represents the energy consumption of the EVs deployed on the roads.

The sampling gathers information contained in the technical specification sheets of the manufacturers. In most cases, the value of the energy consumption is shown specifically in the specification. For the cases in which the manufacturer does not provide the energy consumption value directly in the technical specifications of the vehicle, the consumption is calculated by the ratio between the range that can be achieved with one full charge of the battery installed and the energy of this battery.

On the other hand, the other parameter necessary to estimate the total electric consumption of a vehicle is the driven distance. In this case, annual values are considered in order to stablish a time period which provides representative results.

In order to obtain the annual value of the driven distance, information provided by transport agencies and government department/ministry has been analysed.

Hereunder, the following sub-section estimates the consumption and the average driven distance for each vehicle type considered.

2.2.1 Cars

The cars comprise a largest part of the Spanish vehicle fleet, so the sample of models is also larger than for the other cases. A total of 20 models in the current market from 13 manufacturers are used to obtain the specific consumption for electric cars.

The values of electric consumption obtained from catalogues and technical specification are calculated by the manufacturers according to the conditions defined by EU law. The Worldwide Harmonized Light Vehicle Test Procedure (WLTP) laboratory test has been used to measure the consumption for passenger cars.

- Nissan Leaf: 14.8 kWh/100 km.
- Nissan Leaf e+: 16.4 kWh/100 km.
- Hyundai Kona electric (136 cv): 13.6 kWh/100 km.
- Hyundai Kona electric (204 cv): 14.3 kWh/100 km.
- Hyundai Ioniq electric: 13.6 kWh/100 km.
- Jaguar I-Pace SUV: 21.7 kWh/100 km.
- Volkswagen e-Golf: 15.4 kWh/100 km.
- Tesla model S: 19.7 kWh/100 km.

- Tesla model X: 16.4 kWh/100 km.
- Tesla model 3 (Battery: 75 kWh): 13.4 kWh/100 km.
- Tesla model 3 (Battery: 50 kWh): 12.2 kWh/100 km.
- Renault ZOE: 13.5 kWh/100 km.
- BMW i3s: 14.6 kWh/100 km.
- BMW i3: 13.1 kWh/100 km.
- Peugeot e-208: 14.7 kWh/100 km.
- Mazda MX-30: 17.0 kWh/100 km.
- Kia e-Niro: 15.6 kWh/100 km.
- Mercedes EQC 400 4MATIC: 22.5 kWh/100 km.
- Opel Corsa E: 17.0 kWh/100 km.
- Porsche Taycan 4S: 24.6 kWh/100 km.

After analysing the most recent cars deployed in the Spanish market, the energy consumption average value is 16.2 kWh/100 km.

Once the consumption per km is stablished, the driven distance by car is the other parameter needed for the calculation of the total electric consumption of the vehicle. For the case of the cars, the National Institute of Statistics (INE[1]) provides a value of 12,500 km per year. Additionally, the Department for Transport of Great Britain [1] provides a value of 13,350 km per year in UK and an assessment published in ODYSSEE-MURE[2] estimates that the driven distance by the car in Spain is 12,535 km per year.

Therefore, the average value considered for the driven distance by car of the Spanish fleet is 12,500 km.

2.2.2 Vans

The vans are considered as the most important light duty vehicle in the small-scale commercial trades in the daily life. Moreover, with the increase of the messaging and parcel services promoted by internet and large companies whose online sale platforms have been rising in the recent years, the number of vans will be increased in the next years.

This type of vehicle is usually used for the transport by road of non-heavy loads and it can be found both in the cities and on interurban roads so the driven distance estimated for cars cannot be used for vans due to the difference between the purpose and the use of both vehicles.

Following the same methodology as before, a total of 8 models in the market from 6 manufactures are used to estimate the electric consumption for electric vans.

- Renault Kangoo Z.E.: 15.5 kWh/100 km
- Mercedes e-Vito: 22.7 kWh/100 km

[1] http://www.ine.es/.

[2] https://www.odyssee-mure.eu/publications/efficiency-by-sector/transport/distance-travelled-by-car.html.

- Mercedes EQV: 27.0 kWh/100 km
- Renault Master Z.E.: 27.5 kWh/100 km
- Nissan e-NV200: 20.0 kWh/100 km
- Peugeot partner Electric: 17.7 kWh/100 km
- LDV EV80: 29.0 kWh/100 km
- Volkswagen e-Craft: 21.5 kWh/100 km.

The assessment of the current electric vans market shows that there are not many options for this electric vehicle in Spain when someone wants to get one. With only a sample of 8 different vehicle, the energy consumption average value is 22.6 kWh/100 km.

The other parameter necessary to estimate the total electric consumption is the driven distance by vans. For this type of vehicle, U.S. Department of Transportation[3] provides the values of 18,440 km and 18,570 km for light duty vehicles (short wheelbase and long wheelbase, respectively) and average value of 18,480 km. Furthermore, the Department for Transport of Great Britain provides a value of 20,900 km per year in UK.

Hence, the average value considered for the driven distance by vans is 19,500 km.

2.2.3 Buses

Without considering cars, the buses are the vehicle which the transport of people by road is based in. Buses are integrated in the current life in every term like travel, shopping, work, leisure, etc. both people who live in urban and rural areas. As the main option in the public transport in Spanish cities, the electrification of this vehicle implies great savings in fuel combustion and pollutant exhaust emissions.

In the same way as performed for cars and vans, a sample of a total of 13 buses of 8 manufactures estimate the electric consumption per km for electric buses:

- BYD 35′ Double decker electric bus: 184.1 kWh/100 km
- BYD 30′ Electric transit bus: 143.3 kWh/100 km
- BYD 60′ Electric transit bus: 202.7 kWh/100 km
- BYD 35′ Electric Motor Coach: 156.5 kWh/100 km
- BYD 23′ Electric Motor Coach: 96.8 kWh/100 km
- VDL SLF 120 electric: 154.0 kWh/100 km
- SOR EBN 11: 114.7 kWh/100 km
- AMZ CitySmile 10E: 135.3 kWh/100 km
- Rampini E120: 172.5 kWh/100 km
- Škoda PERUN HE: 146.7 kWh/100 km
- Chariot Motors 12 M: 95.0 kWh/100 km
- Volvo 7900 12 m: 121.0 kWh/100 km
- Volvo 7900 18 m: 161.3 kWh/100 km.

[3]https://www.fhwa.dot.gov/policyinformation/statistics/2018/vm1.cfm.

The analysis of the electric buses in the market provides a value of electric consumption per km equal to 144.9 kWh/100 km.

Buses are vehicles that cover long distances over the day in stablished routes, certain travels, scheduled events, etc., so the driven distance is larger than the case of cars and vans due to the more frequent and continuous use of this vehicle. U.S. Department of Transportation[4] estimates an annual value of 54,700 km and the Department for Transport of Great Britain [2] provides an annual value of 57,000 km.

Both values are similar so the driven distance in this assessment is established every 55,000 km per year.

2.2.4 Motorbikes

Motorbikes have become the main alternative of the cars for many people who live in big cities with a lot of traffic and need to move fast through the street and save traffic jam. Motorbikes offer the possibility of driving among the cars and parking in certain enable places which are exclusive for these vehicles and that makes them a good option for short distance in urban areas or between municipalities near big cities. Also, this vehicle is used on the roads for travelling and for fun so there is another amount of motorbikes which satisfy this need.

It is important to consider that motorbikes engine-cylinder range is very wide, from less than 75 cc to more than 750 cc so this assessment divides the motorbikes into two big groups: Motorbikes with a cubic capacity lower than 125 cc (mainly scooters) and motorbikes with a cubic capacity higher than 125 cc. DGT [1] data shows that in the Spanish regions considered there are 3,166,800 motorbikes (about the 10.6% of the fleet) among them 1,389,234 motorbikes (43.9%) have an engine-cylinder lower than 125 cc and 1,777,566 motorbikes with a cubic capacity higher than 125 cc.

As performed for the previous type of vehicles, the electric consumption per motorbike is calculated by a sample of different models in the current market for both groups. Then, the driven distance is also estimated in the same way.

Motorbikes Engine-Cylinder $\leq$125 cc

A total of 10 motorbikes of 8 manufactures have been used to estimate the electric consumption per km for electric motorbikes:

- Vespa Elettrica L1: 4.2 kWh/100 km
- Gogoro Smartscooter 2 Plus: 3.8 kWh/100 km
- Gigabike Groove: 1.7 kWh/100 km
- Niu n-Serie: 2.3 kWh/100 km
- Niu m-Serie: 1.9 kWh/100 km

[4] https://www.bts.gov/content/bus-fuel-consumption-and-travel-metric.

- Lifan E3: 2.9 kWh/100 km
- Torrot Muvi: 3.1 kWh/100 km
- Eccity 125: 4.2 kWh/100 km
- Bereco cable 3000 WG-3: 4.0 kWh/100 km
- Bereco cable 3000 WL-3: 3.4 kWh/100 km.

Once the electric motorbikes have been analysed, the assessment of the models in the current market provides a value of the electric consumption per km of 3.1 kWh/100 km.

As mentioned before, motorbikes are vehicles used for several purpose in which going to the work or displacement inside the city or from close municipalities are considered as the most important. The strategic plan for traffic safety of motorbikes performed by DGT [3] provides an annual value of driven distance of 11,000 km per vehicle.

Motorbikes Engine-Cylinder >125 cc

A total of 10 motorbikes of 7 manufactures have been used to estimate the electric consumption per km for electric motorbikes:

- Brutus v9 8.6 kWh/100 km
- Johammer J1 6.4 kWh/100 km
- Zero S Z7.2 7.4 kWh/100 km
- Zero SR ZF14.4 7.5 kWh/100 km
- Zero FXS Z11 7.8 kWh/100 km
- Lightning LS-218-12 6.8 kWh/100 km
- Lightning LS-218-20 7.3 kWh/100 km
- Lightning Strike 8.2 kWh/100 km
- Energica EVA 5.9 kWh/100 km
- Harley Davidson Livewire 9.8 kWh/100 km.

These 10 models of the current motorbike market provide an average value of electric specific consumption of 7.6 kWh/100 km.

For the assessment of the driven distance by motorbikes with an engine-cylinder higher than 125 cc is considered that the value of this parameter is the same that of the other group of motorbikes (≤125 cc) so the driven distance for this case is 11,000 km too.

2.2.5 Trucks

Trucks are the main actors of the load transport by road both conveyed load weight and travelled distance. The flexibility provided by this vehicle, regarding routes and schedules, avoid the restriction on timetable and stablished ways which trains have, allowing the customer to optimize costs and time. For this reason, trucks have been stablished themselves as the main option for heavy load transport by road and, for this reason, many project and research have been focused on improving the technology associated to this vehicle.

The trend of the development related to new engines for future trucks is not clear since several strands have been followed during the recent years. The two main strands are the use of solid fuel for the propulsion and the electrification of the vehicles. However, the risks associated to the use of solid fuels seem to tip the balance in favour of the electric trucks.

With the aim to estimate the electric consumption of electric trucks, the same procedure is used as for the other vehicles. A total of 11 models in the current market from 7 manufacturers are used to obtain the specific consumption for electric trucks:

- Mitshubishi eCanter 82.8 kWh/100 km
- BYD T7 70.0 kWh/100 km
- Freightliner eCascadia 136.8 kWh/100 km
- Freightliner eM 106 87.8 kWh/100 km
- Renault D Wide Z.E. 100.0 kWh/100 km
- eMoss EMS 712 75.0 kWh/100 km
- eMoss EMS1008 80.0 kWh/100 km
- eMoss EMS 1620 95.2 kWh/100 km
- eMoss EMS 1824 104.3 kWh/100 km
- Mercedes-Benz eActros 120.0 kWh/100 km
- Tesla Semi 113.9 kWh/100 km.

An average value of 96.9 kWh/100 km is the result of the search in the market of the current electric trucks considering a wide combination of maximum load and size of the models.

Since the trucks are the main transport vehicle by road, they usually drive long distance during their service life. Ministerio de Fomento [4] of the Spanish Government provides different values of driven distance depending on the type of truck: 120,000 km by general articulated cargo trucks, 95,000 km by 3-axis cargo trucks and 90,000 by 2-axis cargo trucks. Furthermore, U.S. Department of Energy [5] and U.S. Department of Transport3 establish the driven distance in 109,685 km and 101,000 km, respectively. Hence, it is assumed that the annual average distance for this assessment is 100,000 km/year.

3 Results

This section shows the main results obtained from the gathered data in previous paragraphs. First of all, the new electric consumption is calculated in the two first point considering different penetration scenarios. Hereunder, the use of the vehicles defines the charging schemes of the electric fleet and the optimal charging curves are obtained. Subsequently, the impact on the grid is analysed considering the increase of the electric consumption and finally, the environmental impact is estimated related to CO_2 emissions savings.

3.1 Equivalent Fleet

The analysis of the current market of the electro vehicles considered in this chapter provided a typical value of the electric consumption per km and annual average driven distance for every vehicle type. However, these values given separately do not show the real impact of the EV insertion in the infrastructure so in this section, a new equivalent fleet is defined in order to offer more visibility in the electricity demand estimation.

With the aim to compare each vehicle type and to simplify futures calculation, a new parameter is defined based on the annual electric consumption of one car named Equivalent Electro-Vehicle (EV). With the utilization of this parameter a new fleet is calculated by transformation of the real buses, vans, trucks, and motorcycle in virtual cars. This way, taking the value of the electric consumption of one car calculated before (16.2 kWh/100 km) and the annual average driven distance (12.500 km/year/car) establish that 1 $EV_{Eq} = 2.02$ MWh/car/year.

For example, in the case of vans, a typical one drive an average distance value of 19,500 km/year and the specific electric consumption have been stablished in 22,6kWh/100 km in previous section so, combining these parameters, the annual electric demand of a van arises to 9.66 MWh/year which, compares with the consumption of a car, is equal to 2.17 cars or 2.17 EV_{Eq}. Therefore, counting the vans of the entire Spanish fleet (2,193,230) and applying the new parameter, the part of equivalent fleet associated with vans is composed of 4,775,290 EV_{Eq}.

Table 2 gives a summary of the annual average driven distance, the specific electric consumption, the annual average consumption, equivalent vehicle, the number of vehicles in the fleet, and the number of vehicles in the equivalent fleet for cars, vans, buses, motorbikes (engine-cylinder $\leq$125 cc and engine-cylinder >125 cc) and trucks.

The result of the transformation of the entire fleet to the equivalent fleet shows that the electrification of all the types considered corresponds to a total of 137,762,037 EV_{Eq} which would be equal to a fleet of that quantity of electric cars. The inclusion of vans, buses, motorbikes, and trucks in this assessment offers a significant difference from the case of the exclusive electrification of cars.

As could have been supposed in the analysis of each vehicle type, trucks have the highest electric consumption and driven distance far from the other vehicles which, in addition of about 2 million of them, represent the major part of the equivalent vehicle fleet. In particular, the number of EV_{Eq} for trucks is 4 time the amount of EV_{Eq} of the other vehicles together.

As a consequence of this result, two scenarios are defined in order to make a difference between the electrification of the entire fleet composed by a total of 137,762,037

Table 2 Equivalent fleet

	Cars	Vans	Buses	Motorbikes	Scooters	Trucks
Average annual travelled distance (km/year)	12500	19500	55000	11000	11000	100000
Specific consumption (kWh/100 km)	16.20	22.61	144.91	7.55	3.14	96.90
Average annual consumption (MWh/Year)	2.02	4.41	79.70	0.83	0.35	96.90
Equivalent vehicles	1.00	2.18	39.36	0.41	0.17	47.86
Fleet	22,113,723	2,193,230	56,071	1,389,234	1,777,566	2,252,425
Equivalent fleet	22,113,723	4,775,291	2,207,078	570,174	303,097	107,792,675

EV_{Eq} and the only electrification of cars, vas, buses, and motorbikes without trucks with an equivalent fleet of 29,969,362 EV_{Eq}.

3.2 *Penetration Scenarios*

Spain is a country where the deployment of the EV is still in process in a early stage regarding both the development of a charging infrastructure and the sales of this type of vehicles. Although the use of hybrid vehicles is fairly common, especially in public transport, the amount of pure electric vehicles is negligible compared to the real Spanish fleet.

The insertion of the EV will not be an instantaneous transition from the current situation to a new electric scenario so the temporal horizons are usually established by the middle of the century. However, since this change is already started in the developed country, some cases are defined depending on the percentage of the total vehicles which are electrified with the aim to simulate a progressive adaptation to the vehicle electrification.

Moreover, the different types of electric vehicles considered in this chapter are analysed separately since, as can be seen in the previous section, trucks represent the major part of the equivalent fleet and, considering each type for their own, the assessment provides a better overview of how the impact on the demand is affected by both the percentage of the electrification and the type of the vehicle.

With the aim to make a proper evaluation, five cases are defined for cars, vans, buses, motorbikes (both cylinder), and trucks. The percentage used for each case are 10, 20, 30, 50, and 100% of the fleet.

Table 3 EV demand

Vehicle	Fleet (Nº vehicles)	Annual electric demand (TWh)				
		10%	20%	30%	50%	100%
Cars	22,113,723	4.48	8.96	13.43	22.39	44.78
Vans	2,193,230	0.97	1.93	2.90	4.83	9.67
Buses	56,071	0.45	0.89	1.34	2.23	4.47
Motorbikes	1,389,234	0.12	0.23	0.35	0.58	1.15
Scooters	1,777,566	0.06	0.12	0.18	0.31	0.61
Trucks	2,252,425	21.83	43.65	65.48	109.13	218.26
Total	29,782,249	27.89	55.79	83.68	139.47	278.95
Total (Without trucks)	27,529,824	6.07	12.14	18.20	30.34	60.68

Table 3 shows the energy demand for all the cases defined above for each vehicle including a final row with the total values.

As can be seen, there is a big difference between the case in which trucks are considered and the case in which only light duty vehicles are electrified, getting in the worst case (with trucks) a value of the total consumption of 279 TWh for the case of a 100% of the electrification.

An interesting result is the comparison between the electric demand due to EV and the current electric demand in Spain, 234 TWh according to the reports of the Spanish Transmission System Operator, REE [6]. If the trucks are electrified with the other vehicles and the inclusion of the electric vehicle if fully deployed, the total electric demand would arise to 513 TWh what exceeds all the expectation for future consumption. Moreover, if it happened, a new energy transportation infrastructure would be needed to support this increase.

Nevertheless, the electrification of only cars, vans, buses and motorbikes would add about 61 TWh (26% of the current demand) being possible to be managed with the current infrastructure. Besides, this result is in line with general estimations which set the increase of the consumption due to the electric vehicle in about a third part of the current demand.

3.3 *Use and Charge of Vehicles*

In general, the electrification of the vehicle will not change the utilization habits of them since they meet the same needs that the current fleet. However, the new schemes of charging will adapt the refuelling system both places to make the charge and the time in which this process is performed. Consequently, these changes also affect the electric energy supply and the hourly demand of electricity switch due to the new demand to be matched.

The new places where the users recharge their vehicles are directly related to the time when it is made. It is due to the new way of charging adds flexibility to adapt the vehicle use to the user requirements without modifying the routes and the expected time of the displacements making easier the use of the vehicles. Moreover, new electric charge stations replace the deployed gas stations, so the current schemes of refuelling also remain in case in which charge on road is necessary, for example in long travels.

In addition, the most important factor which change the current period of refuelling is the availability and cost of the electric energy supply. People obtain electricity from the grid to charge the battery of their vehicles when the price of the energy is the lowest while they are in a place where the vehicle can be connected to the energy system. For example, night fees usually offer lower prices than standard ones and it is a good moment to connect common cars in the houses. Conversely, trucks could not take this advantage due to their use during the night, so the planification of the routes to be achieved is an important issue to be managed in the cost optimization by the companies. Furthermore, some companies could provide free energy to their employees so they would adapt their vehicle charges to the period in which they are working in order to save the cost of that electric energy as well.

Zem2all project [7] estimate the use of electric vehicle by the inclusion of a total of 200 electric vehicles, 220 standard charge points and 23 fast charge point in the Spanish city of Malaga and monitoring all of them. Zem2All provides electric vehicle both particular and commercial user to gather data from different sources. The main results regarding the use of the deployed fleet show that a 20% of the charges are out of the bases (on road), the most of particular users connect their vehicles to the grid after 9 pm, as could be expected, and the commercial users mainly charge the vehicles in rush hour.

Due to all the circumstances mentioned above and more factors not considered in this moment cause of the low insertion of the electric vehicle which could rise when the electrification of the fleet is completed, the estimation of average hourly charge for electric vehicles is supposed to not be reliable.

However, it can be located the period of time during the day which optimize the use of the grid depending on the current electric demand of a typical day to establish the lowest fluctuation in consumption over the 24 h of a day.

The consumption profile for a typical day is obtained applying the Spanish consumption to the electric demand curves provided by REE [6] for all the hours of the year. Then, a standard day is calculated by the hourly average for every day. Figure 2 shows the hourly distribution of the electric demand for a Spanish typical day.

This graph shows that the hours in which the consumption is higher is during the day until about 21:00 overlapping with the usual working hours and the daily life activities for a standard people. This result is suitable for a night charge of the electric vehicles without requiring an additional energy infrastructure. However, there is specifically a peak of consumption at 13:00 which could imply possible grid problems if the electric demand arose at this time due to electric vehicles.

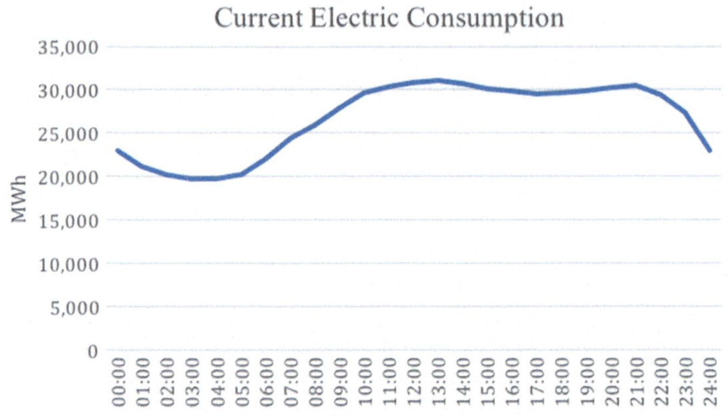

Fig. 2 Current hourly electric consumption

With the aim to draw the optimal charging scheme, it is necessary to estimate how much electricity energy is required for the electric fleet during a typical day and allocate that quantity during the day seeking a constant value of the sum between current and electric vehicle demand.

The estimation of the daily consumption of the electric fleet is calculated considering the average value of the total Spanish demand divided among the 365 days of a year, assuming that all the days the driven distance are equal. This way, for the case of the entire fleet the average daily consumption is 764,234 MWh and without trucks, the value is 166,255 MWh.

In Figs. 3 and 4 the current electric demand is complete with the EV demand for both considered cases (with and without trucks) making equal the total electric consumption in every hours of the day.

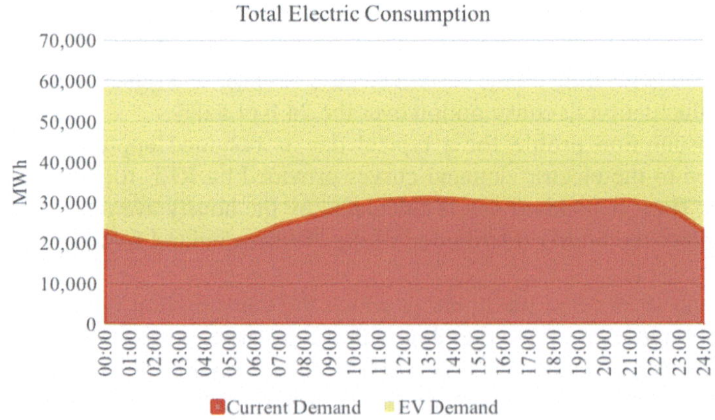

Fig. 3 Hourly total consumption

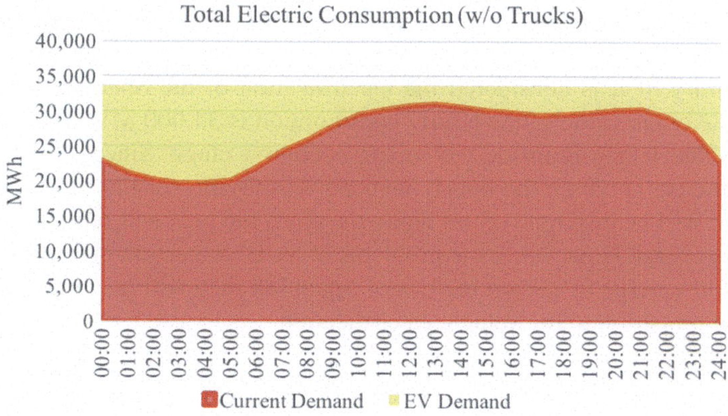

Fig. 4 Hourly electric consumption (w/o trucks)

For the case in which the entire fleet is electrified the hourly consumption arises to 58,656 MWh per hour and the new daily demand is 1,407,751 MWh (about a 219% of the current value which is 643,518 MWh). The hourly value is about 89% higher than the peak of supplied energy (31,075 MWh at 13:00). For the cases without trucks the value of hourly electric consumption is 33,741 MWh per hour, which is about a 7% higher than the peak of electric energy, and the new daily demand is 809,774 MWh (126% over the current daily demand).

Nevertheless, this is the estimation of the optimal scheme of charge and, thus, the real behaviour of the real charge process will be different, following the habits of the users and the adaptation to the new technologies.

3.4 Impact on the Grid

The electrification of the Spanish vehicle fleet is linked to an increase of electric energy which the current energy infrastructure has to face. The transmission system is able to resist a moderate increase of the used electric energy but if there was an excessive gain of the consumption in a certain moment, it could produce a failure in the system.

The data provided by REE [6] with the hourly consumption have the highest value in about 39,000 MWh. This value can be used as a reference to establish the limit of a secure operation of the grid and the system will be reliable in hours in which the consumption is lower than this reference.

Considering the charge schemes estimated in Sect. 3.4, for the cases in which the entire fleet is electrified, the average value from the optimal charging process is about 58,000 MWh, value considerably higher than the established limit, so the grid could be a problem for that increase of the demand. Moreover, the analysed scenario is optimal so in standard condition the consumption curve will have more fluctuations along the days and the peaks of necessary energy will be higher what provokes more problems.

On contrary, if it is considered the electrification of the Spanish fleet without trucks, the average value of the hourly consumption is 34,000 MWh in the optimal scenario, value which is inside the secure operation range. Since it is considered the optimal case in which the hourly demand is constant, for the standard case, in which there will be different values along the days, the grid will remain stable over the year in general. However, some problems of local congestion in the transmission system could emerge in certain regions of Spain but they will be solved during the progressive transition to the vehicle electrification.

3.5 Environmental Impact

The electrification of the vehicle is based on the change of the traditional combustion engine to a new electric one, removing, in this way, the CO_2 emissions (along with other exhaust gases) produced by the vehicles. Thus, the main impact on the transformation from the conventional engines of the traditional fleet to this new technology is the saving on the emissions of the fuel combustion both diesel and gasoline engines.

The first step for the calculation of the emission saving is the estimation of the fuel consumed by the entire Spanish fleet during a year. Afterwards, the CO_2 emissions are estimated for the calculated consumption of the vehicles and for the electric demand necessary in case in which the entire Spanish fleet is electrified. Finally, a comparison is made between the current emission due to combustion engines and the emissions produced to obtain the expected electric demand for the new fleet.

The current combustion engines for vehicles is mainly divided into two big technologies. On one hand, gasoline engines have been the traditional design solution since the vehicles began using hydrocarbons extracted from the oil. On the other hand, diesel engine is the other option in the current market which provides a lower consumption for cheaper fuel but with the handicap of being a more expensive technology. Consequently, the average fuel consumption for the vehicles depends on that and the values for each type are shown in Table 4 [8]:

$$\frac{\text{Consumption}_{\text{Electric motorbike}}}{\text{Cosumption}_{\text{Electric scooter}}}$$

Hereafter, the analysed vehicles have to be divided into two groups regarding the combustion technologies mentioned above. This classification is carried out with the data provided by Dirección General de Tráfico [1] as before. DGT data shows that about 99% of the buses and motorbikes and the 96% of trucks have gasoil engines and the other 1% and 4% are included, respectively, into the gasoline and "other" type. The values for fuel consumption are estimated based on a sample which does not cover these special cases, so it is assumed that every buses, motorbikes and trucks use gasoil fuel, result in line with the values provided by [8].

Table 4 Vehicle fuel consumption

Fuel	Cars	Vans	Buses	Motorbikes	Trucks	Scooters
Average gasoline consumption (l/100 lm)	9.2	13.5	–	6.4	–	2.7[a]
Average diesel consumption (l/100 lm)	7.0	11.7	28.0	–	29.5	–

[a]The value for scooters is not provided by [8] so the calculation of the consumption for scooter is calculated using the following ratio
Besides the used parameters in this coefficient are only for electric vehicles, it is assumed than the ratio is equivalent for vehicles with combustion engines installed. It is due to the consumption depends on many specifications of the vehicles (weight, friction, aerodynamic coefficient, transmission system, etc.) not related to the engines so the difference in the needed power and consumed energy motorbikes and scooters is assumed to be the same

Finally, to calculate the total CO_2 emissions by the entire fleet it is necessary to define the quantity of CO_2 produced by the combustion of each fuel. The ministry Ministerio para la Transición Ecológica y el Reto Demográfico [9] of the Spanish Government stablishes the average value of CO_2 emissions per litre of gasoline and gasoil in 2.35 kgCO_2/l and 2.64 kgCO_2/l, respectively.

Table 5 summarizes all the gathered data and shows the CO_2 emissions for each vehicle type and the entire fleet.

The total emissions amount to 217,395,795 and 73,951,948 tons of CO_2 for the cases with and without the trucks, respectively. In order to estimate the environmental impact, only remains the calculation of the emissions produced to supply the electric energy necessary for the electrified fleet.

In this assessment, it is assumed that the electric energy used to match the demand for electric vehicle is produced by generation with gas. The gas generation technology in Spain has an emission rate of 370 gCO_2/kWh [10] which, applied to the demand of the electric fleet 279 TWh with trucks and 61 TWh without trucks, produce a total of 103,209,757 and 22,452,706 tons of CO_2 for each case. Therefore, in the case in

Table 5 CO_2 vehicle emissions

Type of vehicle	Consumption (l/100 km)		Gasoline/Gasoil (vehicles)		Annual driven distance (km)	Emissions (10^6 tCO_2)	
	Gasoline	Diesel	Gasoline	Diesel		Gasoline	Diesel
Cars	9.2	7.0	9,678,097	12,435,626	12,500	26,155,056	28,726,297
Vans	13.5	11.7	371,758	1,821,472	19,500	2,299,835	10,971,017
Buses	–	28.0	–	56,071	55,000	–	2,279,623
Motorbikes	6.4	–	1,389,234	–	11,000	2,298,349	–
Scooters	2.7	–	1,777,566	–	11,000	1,221,772	–
Trucks	–	29.5	–	2,252,425	100,000	–	175,418,859
Total	11.7	9.5	13,216,655	16,565,594	–	31,975,011	217,395,795
Total w/o trucks	11.7	7.7	13,216,655	14,313,169	–	31,975,011	41,976,936

which the entire fleet is electrified the savings in emission arise to about 53% over the total for the current Spanish fleet and, in the case in which trucks are not electrified, the savings are about 70% over the total, so the pollution related to exhaust fumes could be reduced to less than half.

Finally, it has to be pointed out that in the future, when the entire fleet became electric, the development in renewable energies will have achieved the necessary maturity to supply the total consumption of the electric vehicle with only renewable sources and, in this ideal case, the environmental impact of an electric fleet would be zero.

4 Conclusions

The electrification of an entire vehicle fleet is a process that covers several decades until it is completed. However, there will be a point in which all the considered vehicles are electrified provoking a new scenario for the transportation. As mentioned above, about 30 million of vehicles among cars, vans, motorbikes, buses and trucks, have to change from the typical gasoil/gasoline vehicle to the new battery electric ones and the increase in the electric consumption of 278.95 TWh is added to the current demand.

After performing the estimation of that increase in the electric demand, it is found out that the main part of the consumption is due to the electrification of trucks. For the case in which trucks are not considered, the new electric demand is 60.68 TWh, only the 21.8% of the entire fleet. This result shows that the electric technology applied to trucks could not be enough to develop or could not be proper for this type of vehicle, being hydrogen or another fuel/technology more suitable for them.

Nevertheless, it is clear that the inclusion of the electric vehicle in Spain will produce a scenario fully different regarding the current infrastructure both refueling and electric energy transmission. On one hand, gas station will be replaced by new electric station and the number of them will be lesser due to the possibility of charging in many other places like home, work or wherever place with access charging points. On the other hand, the increase of the demand due to EV is about a 26% of the current demand that, along with this 234 TWh, could produce certain problems in the grid.

Both factors have to be managed together since if the new electric demand due to the vehicle charging was at the same time, local congestion problems would appear during some hours along the day. Figure 4 shows the optimal charging scheme (demand without trucks) to remain the grid in stable conditions without fluctuation over the day, but that curve is the ideal case and the actual behaviour of user won't follow this pattern. With the aim to solve part of this problems, the Spanish government could implement incentives to a proper charge of the electric vehicles both particular and commercial cases and it could be researched with the installation of batteries in the charging points to adapt the demand from the critical hours to the optimal ones.

Finally, it is important to point out the environmental impact of the electrification of the Spanish fleet. The saving on CO_2 delivered to the atmosphere amounts to about 50 million of tons of CO_2 which involves a reduction of approximately the 70% for the case in which trucks are not electrified. If the total fleet is considered, the total saving will amount to 115 million of tons of CO_2 (53% over the total consumption).

The inclusion of the EV in Spain have already stared and is going to become increasingly important in the near future so measures have to be applied from the beginning in order to perform a proper deployment of this new technology in the country to manage and solve potential issues that will appear due to the magnitude of the changes.

References

1. http://www.dgt.es/es/seguridad-vial/estadisticas-e-indicadores/parque-vehiculos/
2. https://assets.publishing.service.gov.uk/government/uploads/system/uploads/attachment_data/file/741953/road-traffic-estimates-in-great-britain-2017.pdf
3. http://www.dgt.es/es/seguridad-vial/consejo-superior-de-seguridad-vial/grupos-trabajo/gt-52-motocicletas-y-seguridad-vial.shtml
4. https://apps.fomento.gob.es/CVP/handlers/pdfhandler.ashx?idpub=TTW103
5. https://afdc.energy.gov/data/
6. https://www.ree.es/es
7. https://www.urbener.com/zem2all/
8. Monzón A, Pérez P, Di Ciommo F (2009) La eficiencia energética y ambiental de los modos de transporte en España
9. http://coches.idae.es/consumo-de-carburante-y-emisiones
10. Arcos-Vargas A, Riviere L, Grid parity and carbon footprint: an analysis for residential solar energy in the mediterranean area

EV Recharging Systems: Technological Review and Impact on the Electric System

Francisco Jesús Matas-Díaz, José María Maza-Ortega, and Ángel Arcos-Vargas

Abstract This chapter is devoted to analyze some basic technical concepts required to understand the charging of electrical vehicles (AC and DC currents, wired or inductive charging, energy flows, charging time, etc.). This analysis will reveal that electric vehicles could be charged in different ways, so it would be required some standardization in terms of charging modes and connectors. This standardization must also be extended to the communication protocols between the electrical vehicle and charging station, safety issues and electrical installations. Without any doubt, a massive penetration of electrical vehicles may considerably impact the power system and, particularly, at the distribution level. For this reason, this chapter evaluates different strategies that can be applied to minimize as much as possible this negative impact that the EV charging may create.

1 Introduction

The current stock of electrical vehicles (EV) has overpassed four million units foreseen by 2017 and a steady increase up to 228 million units is expected by 2030 [1]. EVs, either pure electric or plug-in hybrid, are equipped with a battery which provides the energy required by its operation and must be charged by an external power supply. The stored energy in the battery is directly related to the EV range. Batteries of hybrid EV are between 5 and 15 kWh while those within pure electric vehicles go up to 15 and 85 kWh [2]. For this reason, given the individual energy requirement and the expected EV total number, the e-mobility deployment will considerably affect the planning and operation of the power system due to the need of installing charging stations to provide the energy supply to the EV final user.

This chapter gives a general overview of the main technological aspects related to the EV charging stations and their interaction with the power system. For this purpose, first those basic technical aspects defining a charging station are presented. Then, the main normative and standards related to these EV charging stations are presented.

F. J. Matas-Díaz · J. M. Maza-Ortega · Á. Arcos-Vargas (✉)
University of Seville, Seville, Spain
e-mail: aarcos@us.es

A. Arcos-Vargas (ed.), *The Role of the Electric Vehicle in the Energy Transition*, Green Energy and Technology, https://doi.org/10.1007/978-3-030-50633-9_3

Finally, the impact on the power grid is discussed and the possible technical solutions are proposed to mitigate it.

2 Basic Aspects of an EV Charging Station

The EV charging technology is not unique. There are different options depending on the way that the power is supplied from the grid to the vehicle. The used technology has important implications in the EV on-board equipment, the charging time up to full battery charge, and the impact on the power system. For these reasons, it is convenient to classify the charging systems according to the following characteristics:

- **Type of current supplied to the EV**. The power system is based on alternating current (AC) while the batteries operate in direct current (DC). For this reason, it is mandatory to incorporate a power conditioning unit to connect the battery to the grid which is in charge of the AC to DC conversion. In case of a power flow from the grid to the battery (charging operation mode) this power conditioning unit is called rectifier. The EV power supply is AC if the rectifier is installed on-board but DC if the rectifier is an external unit. AC supply is limited to low power rectifiers due to the fact that its size and weight are proportional to its rated power usually.
- **Energy transfer mode**. The conventional EV charging is done with wires (conductive charging) by means of standardized connectors which will be analyzed later. Alternatively, inductive charging mechanisms have also been developed to transfer power to the battery in a wireless manner through a magnetic field [3]. Finally, with the main objective of reducing the waiting time during the charging operation, alternative strategies based on the battery substitution (*battery swapping*) have been proposed [4]. However, a massive deployment of this technology is not expected due to the reduced number of vehicles compatible with this technological solution nowadays.
- **Power flow direction**. Conventional EV chargers are able to establish a power flow from the distribution system to the EV battery acting as rectifiers. Usually, these devices are based on diode bridges as shown in Fig. 1a [5]. The use of this

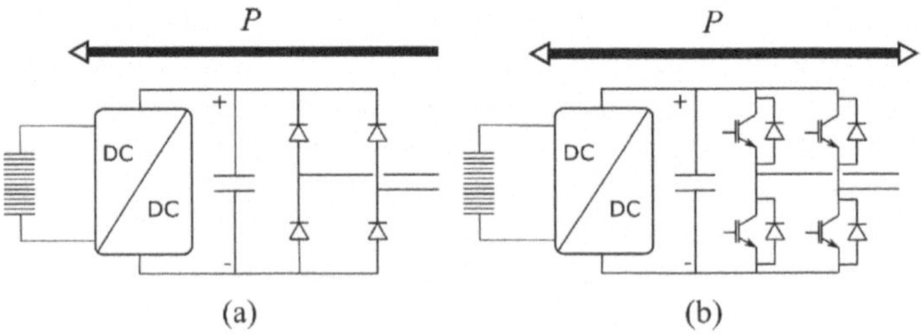

Fig. 1 Single-phase EV chargers: **a** Diode-based EV charger. **b** IGBT-based EV charger

technology has a negative impact on the grid due to the reactive power consumption and the generation of low order harmonics which considerably deteriorates power quality. Alternatively, power converters allowing a bidirectional power flow are also possible. In this case, the power conditioning unit is based on self-commutated switches (MOSFETs or IGBTs) as shown in Fig. 1b. This technology allows inverse power flows, i.e. it is possible to inject active power from the vehicle to the grid. This operation mode is known as vehicle to grid (V2G) which is quite interesting because EVs could be used as a distributed energy storage resource offering ancillary services to the distribution or transmission system [6]. On the other hand, regarding power quality issues, this technology has a reduced content of low order harmonics and almost null reactive power consumption.

- **Charging speed**. This characteristic is conditioned by the power consumption from the grid and the initial and final energy stored in the battery. The charging time will be as lower as higher the power demand, but it has to be considered that this is a nonlinear relationship as shown in Fig. 2. Depending on the active power demand from the network it is possible to establish three charging levels:

  - **Level 1 (slow charge)**. It is the safest and slowest charge but convenient because it can be done using a conventional plug. It uses a 230 V single-phase supply with currents up to 16 A. This is the charge which is used to charge electric motorbikes.
  - **Level 2 (medium or accelerated charge)**. This kind of charge uses higher power demands which in some cases require the use of a three-phase supply with currents up to 63 A.
  - **Level 3 (fast charge)**. It is by far the fastest charging type which demands more active power from the grid. This charge uses DC by means of an external stationary rectifier with voltages about 500 V and controlled currents between 50 and 550 A.

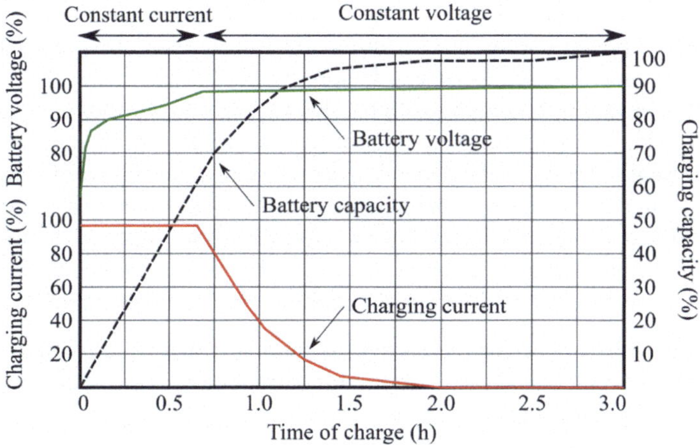

Fig. 2 Evolution of the main magnitudes during an EV battery charging

3 Standardized EV Chargers

Standardization is a key issue to improve the interoperability of charging infrastructure and EV. The standard IEC 61851 [7] copes with all the on-board and external charging devices with rated voltages up to 1.000 V-AC and 1.500 V-DC. This standard defines the classification of the different charging modes as follows:

- **Mode 1**. The charge is done in AC using standard connectors with currents limited to 16 A per phase. In this charging mode, the EV is directly connected to the AC grid without any need of an additional specific device. The maximum active power demanded by this charging mode is 3.7 kW in case of single-phase systems (230 V) and 11 kW in three-phase systems (400 V). In order to assure the adequate protection against indirect contacts, it is required to use a differential protection with earth connection. This charging mode is quite convenient for light vehicles like motorbikes or bicycles but it is forbidden in some countries.
- **Mode 2**. The maximum charging current in this mode is 21 A but it is usually of 16 A per phase. Therefore, the maximum active power is 7.4 kW in case of single-phase systems (230 V) and 22 kW for three-phase systems (400 V). This charging mode has additional functionalities like checking the adequate EV connection to the charging device, checking the earth connection, charging activation, etc. For these reasons, the EV connection cable must incorporate, in addition to the power wires, the corresponding ones for control and safety issues for doing these tasks. This charging mode is quite extended because most of the EV manufacturers include in the EV purchase a charger based on this charging mode.
- **Mode 3**. The charging is done using a specific connector specially designed for EVs. The maximum current associated with this charging mode is 63 A but it is usually of 32 A. Therefore, the maximum active power demand is about 43 kW in case of three-phase systems (400 V). The protection and control functionalities are integrated within the external charging unit. The standard indicates that a pilot control cable is required between the charging device and the EV to assure that the charging process is initiated with the vehicle stopped. This charging mode is being promoted by the European Union, because it allows the EV controlled charging which is fully aligned with the development of the future smart grids.
- **Mode 4**. This charging mode is done in DC by means of an external rectifier which regulates the operation depending on the state of the on-board battery, thus, it requires a communication channel with the vehicle. This mode is exclusively related to fast charging with maximum currents up to 400 A. For this reason, the charging infrastructure is large and expensive in comparison with the other analyzed charging modes. Usually, the fast charging power is limited to 125 kW.

In addition to these charging modes, it is important to highlight that all the connectors between the vehicle and the charging infrastructure are also standardized. The main ones are the following:

- **EEC 7/4 type F (Schuko)**. It is mainly used in Europe for slow charging. For this reason, it is adequate for light vehicles such as motorbikes and bicycles. This connector has a phase, neutral, and ground connections being exclusively applied for single-phase systems with currents below 16 A.

- **SAE J1772**. This connector is similar to the previous one because it includes a phase, neutral and earth connection but it is mainly used in United States. However, it incorporates a communication channel which allows to detect the connectivity between the EV and the charging infrastructure. This type of connect is included in the standard IEC 62196-2 being designated as Type 1 connector. Geographically, it is mainly used in United States and Japan.
- **VDE-AR-E 2623-2-2 (Mennekes)**. It is designed for charging in modes 2 and 3 according to the standard IEC 61851 as previously commented. It is mainly used in Europe, being its design quite similar to SAE J1772. The connector allows the three-phase charge and slow charging. It is included in the standard IEC 62196-2 being designated as Type 2 connector.
- **Scame**. This connector is practically disappearing due to the support of manufacturers to Type 2 connectors. However, it is included in the standard IEC 62196-2 as Type 3 connector. The maximum power is of 22 kW and the connector includes a protection of the connection terminals.
- **CHAdeMO**. It is a connector designed by TEPCO (Tokyo Electric Power Company) specifically suited for the DC charging mode 4 and level 3. Its maximum power is about 62.5 kW and it uses CAN bus as communication system. It is a Type 4 connector according to the standard IEC 62196-2.
- **COMBO (Combined Charging System—CCS)**. It is a modification of SAE J1772 and Mennekes which incorporates a pair of DC wires for allowing the charging modes 2–4.

In addition to the charging modes and connectors, it is possible to find standards for the communication protocols and safety issues:

- **Communication protocols**. The standard ISO 15118 [8] establishes the different communication protocols and the requirements of the physical and data layers. On the other hand, the standard IEC 61851-24 [9] indicates the characteristics of the communication between the vehicle and the charging infrastructure in case of DC charging.
- **Safety issues**. The standard ISO 6469-3 [10] specifies the required protection to avoid electrical hazard for the persons charging the vehicle. Finally, the standard ISO/FDIS 17409 [11] analyzes the safety requirements to take into account for connecting the vehicles to an external charging infrastructure.

Finally, it is also important to standardize all the issues concerning the electrical installations associated with the charging infrastructure to facilitate as much as possible the labor of electrical installers. In this sense, the Spanish ITC BT-52 [12] gives all the details for installing a EV charging station:

- **General requirements of the electrical installation**: rated voltage, neutral connection, earthing system, cable conduits, etc.
- **Connection points**. depending on the charging mode planned for the each EV charging point but always standard connectors are used according to IEC 62196.

- **Protections**. The supply of EV charging points must be designed for guaranteeing the safety of the electrical installation and its users. Therefore, it is required to include the following protections against:
  - **Direct contact**. In order to eliminate the risk of a direct contact with an energized element of the electrical installation adequate insulation and also enclosures or mechanical barriers will be used.
  - **Indirect contact**. For this purpose, any of the following protection systems will be used: automatic disconnection of supply in case of indirect contact (differential switch), devices with insulation class II or isolation transformers.
  - **Overcurrents**. by means of circuit breaker with adequate rated current depending on the charging mode of the charging device.
  - **Overvoltages**. EV chargers must be protected against temporary and transient overvoltages. The former ones are usually caused by the neutral wire breakdown while the later ones happen due to atmospheric phenomena or short-circuit faults.
  - **External agents**, particularly water, penetration of solid foreign content, mechanical impact and corrosion.
- **Energy meters**. Two types of energy meters can be found in the electrical installations of EV charging infrastructure: those used for billing purposes and other auxiliary ones which are not always present but are required for metering the individual EV energy consumption.
- **Installation schemes**. Different installations schemes have been proposed depending on the installation place, the location of energy meters, and the number of charging stations:
  - Collective scheme with a main energy meter at the beginning of the electrical installation is shown in Fig. 3. This scheme is devoted for buildings where the EV chargers are located in the parking space. In this scheme the use of auxiliary energy meters is required for measuring the energy demanded by each individual EV charging point.
  - Individual scheme with a common energy meter for the house and the EV charger as shown in Fig. 4a. Note that in this case just one main energy meter per each house is required. In addition, the fuse installed before the main energy meter must be rated conveniently to cope with the total load (EV charger and the house). It is also possible to follow the installation scheme as shown in Fig. 4b where two independent energy meters for the house and the EV charging point are used.
  - Scheme for individual family-house: In this case the installation is straightforward because a circuit specially devoted is just added to the EV charging point as shown in Fig. 5.

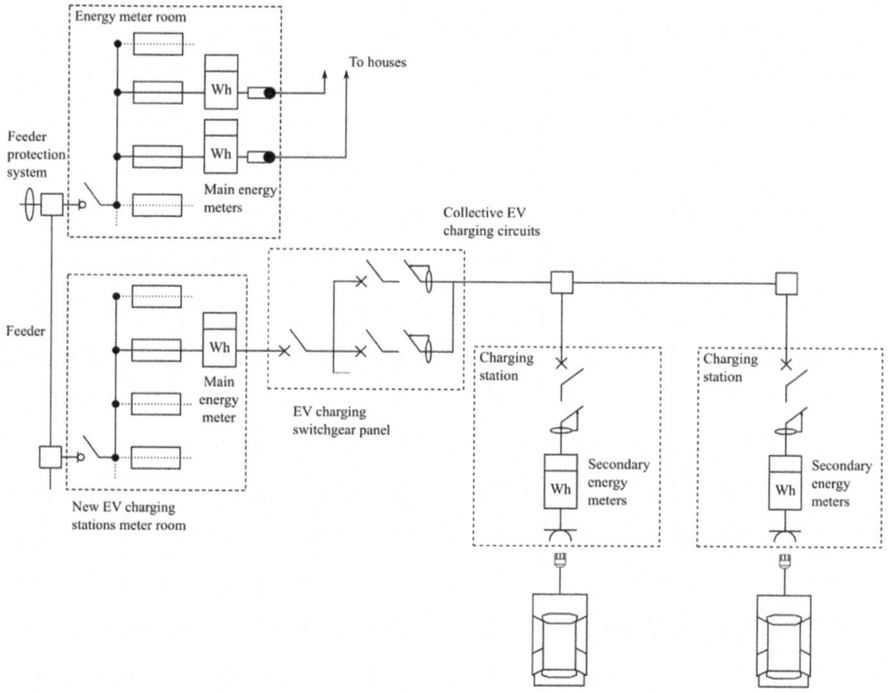

Fig. 3 Collective scheme of different EV charging points in a building. *Source* ITC BT-52 [12]

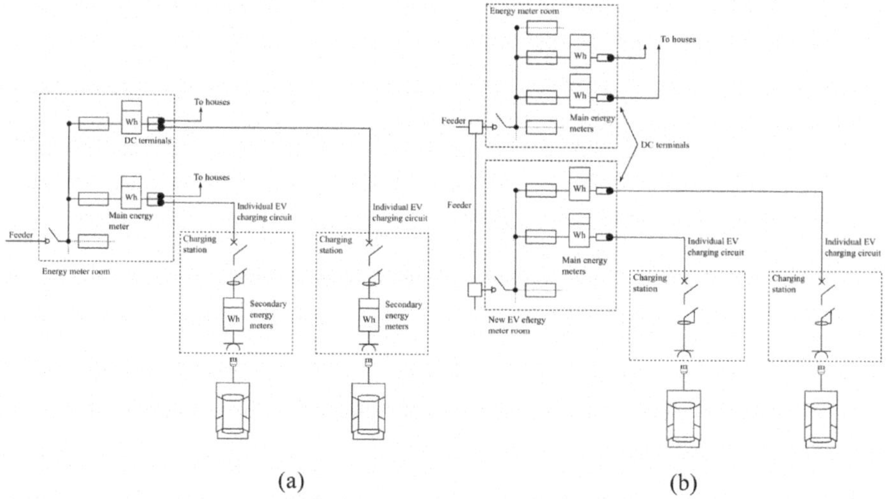

Fig. 4 Individual scheme of different EV charging points in a building: **a** Only one main energy meter. **b** Two energy meters. *Source* ITC BT-52 [12]

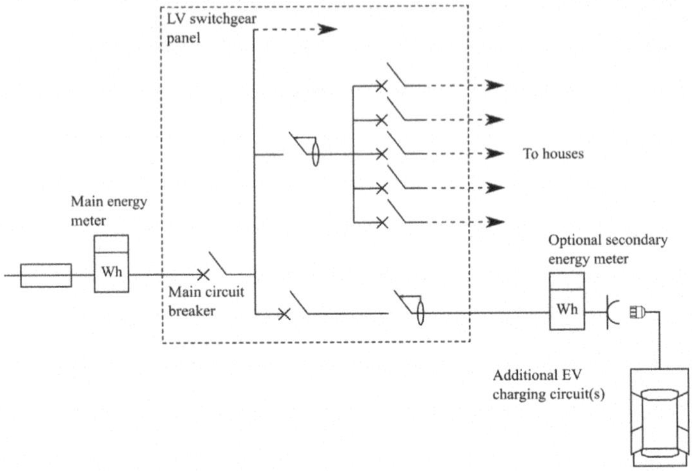

Fig. 5 Individual scheme of different EV charging points in a building. **a** Only one main energy meters. **b** Two energy meters. *Source* ITC BT-52 [12]

1. **Impact on the power grid**

The EV impact on the power grid will mainly depend on the final number of this type of vehicles and their corresponding charging. In spite of the total number of EV is quite reduced nowadays, a steady increase of this technology is expected in the future years. However, the following barriers related to technical, economic, and infrastructure issues should be overcome:

- **Technical barriers**. Probably one of the main concerns related to e-mobility is the EV range. Nowadays, the average of batteries is about 20–30 kWh which allows a range between 150 and 200 km but it is also possible to find some models with 75 kWh batteries to extend the range up to 500 km as shown in Table 1. This reduced range usually creates range anxiety in drivers due to the unavoidable comparison between this new technology and conventional internal combustion engine (ICE) vehicles, which are able to travel longer distances and refuel in a short period of time [13].
- **Economic barriers**. The EV cost is nowadays very high compared to one of the ICE vehicles with similar performance. This is mainly due to the battery cost, usually based on Li-ion technology, which is still quite expensive. However, the actual trend targets a considerable cost reduction because of the maturity of this technology. In this regard, it is important to point out that the reduction cost between 2008 and 2015 has been about 73% from an initial cost of 1.000 $/kWh to just 268 $/kWh [14]. Additionally, most part of prospective analysis hold that the EV competitiveness will happen with a battery cost of about 150 $/kWh. For these reasons, if this forecast is finally verified it should be possible to incorporate larger

Table 1 Current EV range

Model	Range (km)
Tesla Model S	539
Tesla model 3	500
Tesla MOdel X	475
Chevy Bolt	383
Nissan Leaf	350
Renault Zoe	300
Volkswagen e-Golf	300
Hyundai Ioniq	250
BMW i3	200
Kia Soul	200

Source https://evobsession.com/10-electric-cars-range-new/

batteries within the vehicles, increasing the vehicle range and fully removing one of the previously analyzed technical barriers.

- **Barriers related to the charging infrastructure**. Without any doubt, the charging infrastructure is key for the transition towards a fully electrified transportation sector. However, it is a complex problem because it is quite difficult to satisfy at the same time the interest of the EV owners and charging agents during the initial years of the technology deployment (chicken and egg situation). Due to the scarcity of EVs, a massive deployment of charging infrastructure is not undertaken given the fact that the related investment is not profitable enough. Likewise, the absence of charging infrastructure considerably hinders the EV purchase and prevents their use for large routes. For this reason [15], justifies the need of applying economic incentives to the charging infrastructure, at least during the first years up to the massive deployment of this new technology.

The analysis of these barriers clearly evidences that they are not unsurmountable problems and, therefore, a steadily growth of the electric vehicle fleet is expected as long as their cost gets reduced and their range gets increased which is clearly related to the battery development. For this reason, e-mobility is expected to affect the power grid in a global manner at the transmission system level and also locally in the distribution systems where the charging infrastructure will be connected to.

On the one hand, considering a global point of view, in 2020 an EV fleet of about 500.000 units is expected in Spain which may represent an additional consumption of about 1,5 TWh, just a 0.49% of additional consumption considering that the total demand is about 303,901 TWh [16]. Though this figure is not so large, it has to be considered that if the total national vehicle fleet, about 22 million units, is substituted by its corresponding EV counterpart, the consumption will grow up to 66 TWh, representing 22% of the annual demand. The affection to the power system is not only reduced to an increase of energy demand but also it has to be considered the modification of demand curves depending on the way the EV charge is performed.

An uncontrolled EV charging may produce an increase of the peak power [17] with an increment of the system operation cost [18].

On the other hand, considering the local impact on the distribution systems, it is possible to confirm that the penetration of EV chargers leads to a load growth of the medium and low voltage grids. The consequences of this load increment are clear [19]:

- Voltage drop increases due to the load growth. It is possible to face undervoltages in those instants where simultaneous EV charges happen. This may negatively affect the power quality perceived by the final user which, on the other hand, is regulated by stringent standards [20].
- Line congestion because of loading above the thermal limits (ampacity limits) in case of massive simultaneous EV charging.
- Increase of active power losses due to the load growth.
- Reduction of the transformer useful life due to the load growth [21].
- Increase of voltage and current unbalance. Low power chargers are usually single-phase devices as outlined in a previous section. In this sense, it is important to equally distribute the different single-phase chargers between the existing phases to achieve a load balance. Otherwise, an unbalance operation of the system may considerably increase the power losses [22]. Additionally, the operation with unbalance load considerably deteriorates the power quality because of the increase of the unbalance voltage indices which are limited by standards [20].
- Finally, it has to be considered that all the EV chargers are based on power electronic components which are devoted to adapt the AC grid with the DC battery. This kind of devices are nonlinear loads meaning that they absorb a non-sinusoidal current even with a sinusoidal voltage supply. The demanded current, without the adequate countermeasures, may have a high harmonic content which negatively affect the power quality of the distribution system [23].

For all these reasons, it is required to look for alternative EV charging strategies aiming at reducing these negative impacts in case of a massive deployment of e-mobility.

2. Strategies for minimizing the impact on the power grid

In order to minimize as much as possible the negative impact that EV charging may have in the power system it is required to implement smart charging strategies. These strategies can be classified according to the manner that the additional load related to the EV charging is managed:

- **Classical network reinforcement**. This is the most basic way of solving the problem caused by the additional load associated to the EV charge. This strategy assumes that the EV is an uncontrollable load and, for this reason, it is based on new investments on classical network assets (lines and transformers) to avoid network congestions. Following this line it has been proposed new planning models considering the load growth due to the EV [24]. However, it has to be considered that this network reinforcement is not always a straightforward task given the fact that in densely populated areas the distribution network is mainly underground. In these cases, it is required to open ditches along the streets which may cause severe inconveniences (temporal street closure with a traffic reordering, noise during the works, etc.) [25]. For these reasons, considering the maturity the technologies related to smart grids, it is possible to formulate alternative solutions which are analyzed next.

- **Optimal location of EV charging stations**. The main objective of these strategies is to find out the optimal location and rated power of each EV charging station considering the optimization of a given objective function. There are many approaches to this problem in the specialized literature depending on the selected objective function. For instance, [21] proposes to reduce the system cost while [26] minimizes the investment and operational costs. However, it has to be considered that the optimal planning of distribution networks is a complex task which is usually solved by applying multi-period approaches [27, 28]. Moreover, it is important to consider that the solution may also involve other technologies as energy storage systems or renewable energies which may considerably reduce the impact on the distribution system but affect the overall solution cost [29].
- **Controlled EV charging**. The aim of these algorithms is to share the EV additional load along the day for avoiding the network congestions and, simultaneously, fulfill the requirements of the final EV users with respect to the charge of their corresponding EV batteries. Several algorithms can be found for this purpose, which can be classified according to the implemented control architecture and optimized objective function. Regarding the control architecture, it is possible to find either centralized or decentralized approaches. The former methods compute the optimal demand curves of each EV charging station for minimizing a global operation objective such as minimizing the distribution system power losses. For this purpose, it is required to gather information about the state of the system in real time which requires a complex communication infrastructure. This infrastructure, additionally, can be used for communicating the central control system with the different EV charging stations. In this way, the central controller may send the optimal setpoints to the different EV charging stations which optimize the system operation [30, 31]. However, it has to be considered that, in spite of obtaining a global optimum, this kind of applications rely on a complex and costly communication infrastructure which may fail [32]. Additionally, the computational complexity of the mathematical problem can be high in case of a massive deployment of EV charging stations [33]. For these reasons, decentralized control algorithms can be an alternative solution to the traditional centralized approach. These kinds of algorithms distribute the decision making between independent controllers of different EV charging stations. Different implementations can be found in the specialized literature ranging from simple algorithms with operation based just on local measurements [34] to more complex controllers with some communication capabilities which define its operating strategies through the information shared with their local neighbors [33, 35]. Regarding the different

objective functions used in controlled EV charging, it can be found either technical and economic criteria. On the one hand, the usual technical objectives are to maintain the voltages within the statutory limits [34] and also to minimize the active power losses [31]. On the other hand, it has been proposed different economic criteria depending on the different stakeholders involved in the EV charging. Thus, [30] proposes to minimize the operational cost of the utility, [35] evaluates the operation for maximizing the profit of EV charging agent and [36] proposes an algorithm for minimizing the EV charging cost for the final user.

- **Time of use (ToU) electricity tariffs**. The main objective of these strategies is to establish a policy of EV charging cost through an electricity tariff with variable prices along the day. In this way, the final EV user may have an economic signal for transferring the consumption to those periods with lower prices [37]. Those periods with an expected high loading and with a high probability of network congestion may have higher prices than other periods with lower loads. Therefore, the EV final user is persuaded to displace the EV charging to these low-price periods preventing the network congestion.
- **Use of distributed renewable generation**. The objective of these strategies is to diminish the active power demand and the energy consumption from the grid during the EV charging by means of a local generation contribution [38]. This generation could be integrated in the EV charging station or close to it for, in this way, reducing the EV charging impact in the grid. However, this solution is difficult to apply in urban areas where the available space is sometimes reduced.
- **Use of energy storage systems**. This can be an alternative to the use of renewable generation in case of urban areas to reduce the risk of network congestion. The objective of this strategy is to flatten as much as possible the demand curve of the EV charging station [39]. In this way, the peak power of the EV charging station gets reduced preventing the network congestion. However, it has to be considered that the use of energy storage systems does not reduce the energy consumption from the grid but just distribute it along the day.
- **Sharing of EV charge between different feeders**. All the aforementioned methods for reducing the EV charging demand are quite interesting for the distribution system operator because of the investment deferral but some of them presents some problems. On the one hand, the use of renewable energies in EV charging stations located in urban networks is questionable due to the space requirements for achieving the required relevant power to reduce the impact on the grid. On the other hand, the current high costs associated with the energy storage systems pose a clear economic barrier which hinders its use for this purpose. However, it is possible to propose an alternative approach which perfectly fits to the characteristics of urban distribution systems. Note that in urban areas it is common to find radial distribution feeders departing from secondary substations and with final nodes quite close each other. The idea is to install the EV charging stations in such a way that they could be fed from several feeders simultaneously as shown in Fig. 6. In this way, it should be possible to share the EV charging load between the different feeders depending on their load [40]. The main goal of this scheme is that it is not required any additional investment compared to

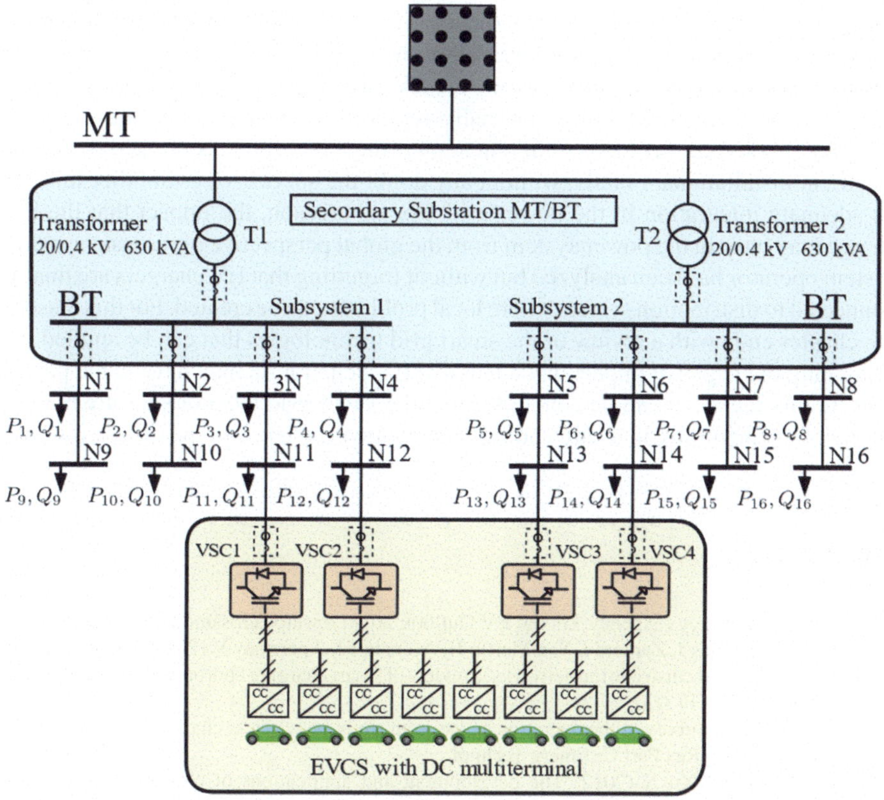

Fig. 6 EV charging station connected to different feeders simultaneously to distribute the load between them. *Source* García-López et al. [40]

a classical EV charging station as its main hardware components are the same. Note that the proposed EV charging station slightly modifies the connections of the AC/DC and DC/DC converters composing a conventional EV charger to create a multiterminal DC link able to control in a suitable manner the EV demand.

4 Conclusions

This chapter has done a review of the technology involved in the EV charging stations which play a significant role in the decarbonization of the transportation sector. The reduction of greenhouse gas emissions and the improvement of the air quality of urban areas is a must nowadays. This will be accomplished only in case of a successful transition towards a new electromobility paradigm. However, it will be required to deploy the adequate EV charging infrastructure to support the steadily growing fleet of electric vehicles. The EV charging depends on several technological issues that have been reviewed in this chapter (energy transfer mode, power flow direction, conversion technology, charging level, etc.). Interoperability has been revealed as a key issue because it is required to assure the EV charging irrespective of the country. For doing so, standardization is crucially required to clearly define the different charging modes, connectors, communication protocols, safety issues, and design of electrical installations. Finally, without any doubt the success of e-mobility falls on its adequate integration in the power grid. For this reason, the impact that the EV charge may have in the power system from the global perspective of the transmission system operator has been analyzed but without forgetting that EV chargers are finally connected to distribution system where local problems can be created. For this reason, the chapter ends with a review of the smart grid technologies that can be applied for mitigating the negative impact that a massive EV penetration may have in the power grid. In this regard, it can be concluded that the smart grid technologies are mature enough for integrating in a safe and reliable manner the EV into the power system.

References

1. International Enregy Agency. Global EV Outlook 2018: towards crossmodal electrification
2. García-Villalobos J, Zamora I, San Martín JI, Asensio FJ, Aperribay V (2014) Plug-in electric vehicles in electric distribution networks: a review of smart charging approaches. Renew Sustain Energy Rev 38:717–731
3. Jang YJ (2018) Survey of the operation and system study on wireless charging electric vehicle systems. Transp Res Part C: Emerg Technol
4. Chen L Wu M, Xu X (2012) The development and applications of charging/battery swap technologies for EVS. In: IEEE 2012 China international conference on electricity distribution, Shanghai, pp 1–7
5. Yilmaz M, Krein PT (2013) Review of battery charger topologies, charging power levels, and infrastructure for plug-in electric and hybrid vehicles. IEEE Trans Power Electron 28(5):2151–2169
6. Kempton W, Udo V, Huber K, Komara K, Letendre S, Baker S, Brunner D, Pearre N (2009) A test of vehicle-to-grid (V2G) for energy storage and frequency regulation in the PJM system
7. IEC 61851-1:2017. Electric vehicle conductive charging system—Part 1: general requirements
8. ISO 15118. Road vehicles—Vehicle to grid communication interface
9. IEC 61851-24. *Electric vehicle conductive charging system - Part 24: Digital communication between a d.c. EV charging station and an electric vehicle for control of DC charging*
10. ISO 6469-3. Electrically propelled road vehicles—Safety specifications. Part 3: protection of persons against electric shock
11. ISO/FDIS 17409. Electrically propelled road vehicles—Connection to an external electric power supply—Safety requirements
12. ITC BT-52. Special installations. EV charging infrastructure (in Spanish)
13. Caruso C (2016) Why range anxiety for electric cars is overblown sustainable energy
14. Grantham Institute (2017) Expect the unexpected. The Disruptive Power of Low-carbon Technology
15. Arcos-Vargas A, Maza-Ortega JM, Núñez F (2018) Proposals for the deployment of the e-mobility: identified barriers and actions to be taken (in Spanish). Real Academia de Ingeniería de España
16. Comisión Nacional de los Mercados y la Competencia (2015) Report on the planning proposal of the Spanish transmission system 2015–2020 (in Spanish)

17. Schill WP, Gerbaulet C (2015) Power system impacts of electric vehicles in Germany: charging with coal or renewables? Appl Energy 156:185–196
18. Hanemann P, Behnert M, Bruckner T (2017) Effects of electric vehicle charging strategies on the German power system. Appl Energy 203:608–622
19. Lopes JAP, Soares FJ, Almeida PMR (2011) Integration of electric vehicles in the electric power system. Proc IEEE 99(1):168–183
20. UNE-EN 50160. Voltage characteristics of electricity supplied by public distribution networks
21. Abdelsamad SF, Morsi WG, Sidhu TS (2016) Optimal secondary distribution system design considering plug-in electric vehicles. Electr Power Syst Res 130:266–276
22. Jiang C, Torquato R, Salles D, Xu W (2014) Method to assess the power-quality impact of plug-in electric vehicles. IEEE Trans Power Deliv 29(2):958–965
23. Gómez J, Morcos M (2003) Impact of EV battery chargers on the power quality of distribution systems. IEEE Trans Power Deliv 18(3):975–981
24. Fernandez LP, San Roman TG, Cossent R, Domingo CM, Frias P (2011) Assessment of the impact of plug-in electric vehicles on distribution networks. IEEE Trans Power Syst 26(1):206–213
25. Gómez-Expósito A. Arcos-Vargas A, Maza-Ortega JM, Rosendo-Macías JA, Álvarez-Cordero G, Carillo-Aparicio S, González-Lara J, Morales-Wagner D, González-García T (2018) City-friendly smart network technologies and infrastructures: the Spanish experience. Proc IEEE 106(4):626–660
26. Simorgh H, Doagou-Mojarrad H, Razmi H, Gharehpetian (2018) Cost-based optimal siting and sizing of electric vehicle charging stations considering demand response programmes. IET Gener Transm Distrib 12(8):1712–1720
27. Xiang Y, Liu J, Liu Y (2016) Optimal active distribution system management considering aggregated plug-in electric vehicles. Electr Power Syst Res 131
28. Xu H, Miao S, Zhang C, Shi D (2013) Optimal placement of charging infrastructures for large-scale integration of pure electric vehicles into grid. Int J Electr Power Energy Syst 53:159–165
29. de Quevedo PM, Muñoz-Delgado G, Contreras J, Impact of electric vehicles on the expansion planning of distribution systems considering renewable energy, storage and charging stations. IEEE Trans Smart Grid. Early Access
30. Arias NB, Franco JF, Lavorato M, Romero R (2017) Metaheuristic optimization algorithms for the optimal coordination of plug-in electric vehicle charging in distribution systems with distributed generation. Electr Power Syst Res 142:351–361
31. Mehmood KK, Khan SU, Lee S, Haider Z, Rafique MK, Kim C (2018) A real-time optimal coordination scheme for the voltage regulation of a distribution network including an OLTC, capacitor banks, and multiple distributed energy resources. Int J Electr Power Energy Syst 94:1–14
32. Marano-Marcolini A, Barragán-Villarejo M, Fragkioudaki A, Maza-Ortega JM, Romero-Ramos E, de la Villa-Jaén A, Carmona-Delgado C (2016) DC link operation in smart distribution systems with communication interruptions. IEEE Trans Smart Grid 7(6):2962–2970
33. Karfopoulos EL, Hatziargyriou ND (2013) A multi-agent system for controlled charging of a large population of electric vehicles. IEEE Trans Power Syst 28(2):1196–1204
34. Cardona JE, López JC, Rider MJ (2018) Decentralized electric vehicles charging coordination using only local voltage magnitude measurements. Electr Power Syst Res 161:139–151
35. Unda IG, Papadopoulos P, Skarvelis-Kazakos S, Cipcigan LM, Jenkins N, Zabala E (2014) Management of electric vehicle battery charging in distribution networks with multi-agent systems. Electr Power Syst Res 110:172–179
36. Hamidi RJ, Livani H (2017) Myopic real-time decentralized charging management of plug-in hybrid electric vehicles. Electr Power Syst Res 143:522–532
37. Cao Y et al (2012) An optimized EV charging model considering TOU price and SOC curve. IEEE Trans Smart Grid 3(1):388–393
38. Mouli GRC, Bauer P, Zeman M (2016) System design for a solar powered electric vehicle charging station for workplaces. Appl Energy 168:434–443

39. Bai S, Lukic SM (2013) Unified active filter and energy storage system for an MW electric vehicle charging station. IEEE Trans Power Electron 28(12):5793–5803
40. García-López FP, Barragán-Villarejo M, Maza-Ortega JM, Expósito AG (2015) Multiterminal electrical charging station for LV networks. IEEE PowerTech (Eindhoven)

Success Factors in EV Deployment: An Economic Analysis

Fernando Núñez and Angel Arcos-Vargas

Abstract More than one million new electric vehicles (EV) were registered worldwide in the year 2017 (record to date), reaching the stock of this type of vehicles three million units in this year. The objective of this paper is to analyse the key factors of the important deployment of the EV from a comparative perspective at international level. The study analyses, using a stochastic frontier model for panel data, the effect on EV registration of the charging infrastructure (fast and slow chargers), the storage technology and the measures to stimulate the supply and demand of this type of vehicles.

1 Introduction

In recent years, the electric vehicle has gone from being a technical curiosity, to an alternative transport valid for most users. This fact is confirmed by the more than one million electric vehicles put into circulation in 2017, reaching a total fleet of more than three million vehicles, having experienced a 50% increase compared to 2016 [4].

This growth can be partly explained by technological improvement and the reduction of associated costs, currently presenting, in many countries, a lower total cost for the ownership in many countries (e.g. in Europe), being therefore an efficient, economic and environmentally friendly alternative [12].

Technological factors may explain the overall evolution, but there are significant differences between countries that are relatively close, such as Norway, with a share of electric vehicles close to 40%, compared to Spain, Italy and Greece, whose share is less than 1%. Figure 1 shows the market share of electric vehicles in different countries.

This article aims to explore the causes that determine these differences between countries, through econometric analysis. To this end, in addition to estimating the effect of the technological variables (cost of the battery and charge density), which can

F. Núñez · A. Arcos-Vargas (✉)
University of Seville, Seville, Spain
e-mail: aarcos@us.es

A. Arcos-Vargas (ed.), *The Role of the Electric Vehicle in the Energy Transition*, Green Energy and Technology, https://doi.org/10.1007/978-3-030-50633-9_4

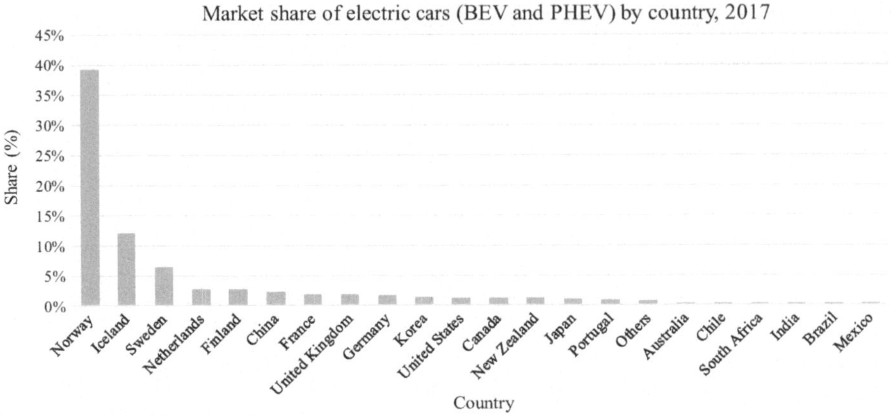

Fig. 1 Participation of electric vehicles in new registrations 2017. *Source* Own elaboration based on IEA data

be assumed to be common to all countries, the provision of recharging infrastructure for each country, which is the variable that most affects users, is included as an explanatory variable when making the decision to purchase an electric vehicle.

The analysis results, in addition to providing an estimate of the impact that a variation in the variables considered would have (cost of the battery, charge density, number of fast chargers and number of slow chargers), presents an analysis of the individual effects of each country, which gives an idea of its level of efficiency with the same infrastructure provision. These individual effects are contrasted with the industrial policies of each country (market promotion and infrastructure promotion), which will allow, in a qualitative way, to make recommendations.

Although the International Energy Agency's report Global EV Outlook [4], with data from 2017, has recently been published, for the econometric analysis the year 2016 has been used, since it was the last year in which Spain and Italy appeared broken down, which, as mentioned above, are countries with a very low level of penetration and interesting for the authors' line of research.

The rest of the chapter is developed as follows: a description of the data considered (Sect. 2) as well as an international comparison in the period considered are presented below. On the basis of these data, Sect. 3 is devoted to a panel data analysis, estimating the fixed effects (determinants of battery vehicle registrations). As well as the efficiency levels of each country (individual effects), which are contrasted with the market promotion and infrastructure policies in place in each country. The work ends with some conclusions and recommendations of industrial policy (Sect. 4), where the effect that a variation of the explanatory variables would have is presented in a quantitative way and, in a qualitative way, the relation of the efficiency of each country with its promotion policies.

2 International Comparison. Data Description

The data published by the International Energy Agency (IEA) in its Global EV Outlook [4] report shows how in 2017 the threshold of one million new electric vehicle (EV) registrations on the road worldwide was exceeded for the first time (Fig. 2), with EV stock exceeding 3 million units in that year. Specifically, 2017 closed with 1,148,700 new registrations, of which 750.5 thousand corresponded to electric battery vehicles (EVs) and 398.2 thousand to hybrid vehicles (HVs). The growth rates in new registrations of both types of vehicles have moderated over the last decade (as the volume of registrations has grown), although these rates are rising in 2017 compared to the previous year. Thus, in 2016 (compared to 2015) the rates were 43.3% and 29.3% for BV and HV, respectively, while in 2017 (compared to 2016) these rates were 60.9% and 40.1%, respectively. These relatively significant growth rates reflect the significant effort that governments and industries in the various countries promoting EV are making in recent years to achieve the electrification of the transport sector. In addition to these growths, the strong growth experienced by fast charging points in recent years must be added. Bear in mind that fast chargers have grown worldwide by 46.8% in 2017 (going from 76.3 thousand units to 112 thousand units between 2016 and 2017), growth that is mainly explained by the strong boost that China, Korea, Germany and the USA are giving to this charging infrastructure. On the other hand, the slow chargers show a more moderate growth in 2017 (up 34.1%), going from 237.2 thousand units in 2016 to 318.1 thousand units in 2017.

Figure 3 shows the percentage distribution of the enrolment of EV, and its two variants (BV and HV), by country in the years 2010 and 2015, which are the extremes

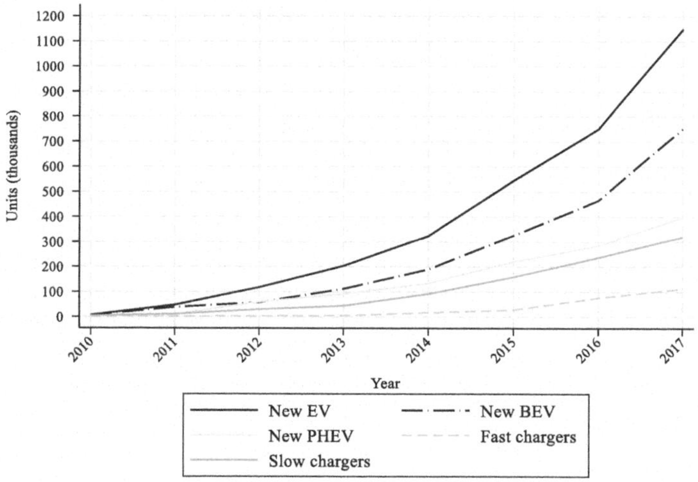

Fig. 2 Global evolution of EV registrations and chargers for EVs. 2010–2016. *Source* Own elaboration based on IEA data

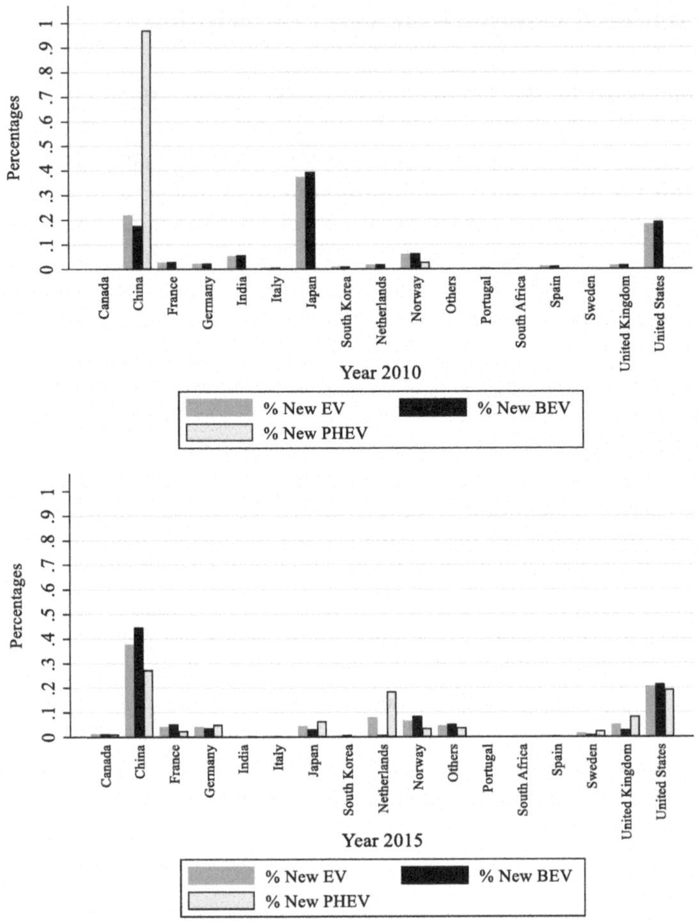

Fig. 3 Global distribution of EV registrations. 2010 versus 2015. *Source* Own elaboration based on IEA data. *Note* the block of 'other countries' consists of Austria, Belgium, Bulgaria, Croatia, Cyprus, Denmark, Estonia, Finland, Greece, Hungary, Iceland, Ireland, Latvia, Liechtenstein, Lithuania, Luxembourg, Malta, Poland, Romania, Slovak Republic, Slovenia and Turkey

of the time interval analysed.[1] In 2010, China accounted for almost all new enrolments of HV. Apart from China, only Norway registered vehicles of this type in 2010, although in a very small volume compared to China. Japan, the United States and China were the countries that dominated the new registrations of WV, with Japan representing almost 40% of total registrations that year; the United States and China were close to 20% of the registrations of this type of vehicle.

[1]The year 2016 has been eliminated from the comparative analysis, both at the descriptive and econometric levels, because we do not have the data for Spain, Italy or Portugal for that year. These countries are part of the group called 'Others' in the Global EV Outlook [5].

Five years later, in 2015, the landscape changes significantly. China substantially decreases its weight in HV enrolment; specifically, its share drops from 97.1 to 27.3%, leaving room for countries such as the USA (19.3%), the Netherlands (18.6%), Japan (6.4%) or the UK (8.3%). Likewise, Japan was the country that experienced the greatest drop in participation in GV registrations (from 39.4% in 2010 to 3.1% in 2015); in its place, China took over as the country with the greatest weight in this type of registrations, with a percentage of 44.6% in 2015 (its weight was 17.6% in 2010). China is followed, in the new registrations of GV, by the USA and Norway, which increase their weight slightly as compared with the year 2010 (going from 19.2% to 21.6% and from 6.3% to 8.4%, respectively).

Therefore, China (followed by the USA) is the leading country in terms of BV and HV registrations in 2015. This country is also currently the world leader in the deployment of e-scooters and electric buses.

Figure 4 depicts the evolution over time of the EV market share in the different countries analysed, a share defined as the percentage of EV purchases over total purchases of all types of vehicles.

Norway and the Netherlands (both represented on the right axis of Fig. 4 for better visualization) have the highest market shares during the whole period; reaching 23% and 10%, respectively, in 2015. In that year, the market share of electric cars exceeds 2% in Sweden and 1% in countries such as China, France or the United Kingdom. The progress of the electric car in all these countries is mainly due to three factors:

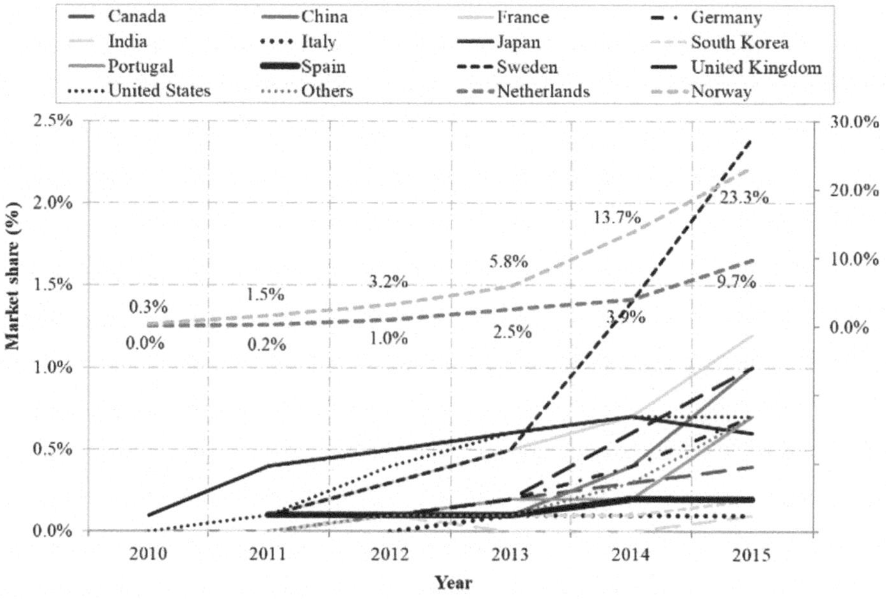

Fig. 4 Market share of EV within each country 2010–2015. *Source* Own elaboration based on IEA data

(1) reduction in the production costs of manufacturing the EV (including the battery),
(2) development of the auxiliary industry (mainly, the one in charge of supplying electric chargers), and
(3) existence of policies to encourage the supply and demand of EV and auxiliary equipment.

Regarding the first determinant, the reduction in the EV's manufacturing costs undoubtedly involves reducing the manufacturing cost of the storage batteries. Battery costs have been reduced by up to 75% since 2008 and are expected to fall further in the coming years. Figure 5 shows how the relationship between the flow of new BV (horizontal axis) and the variables 'battery cost' (measured in US dollars per kWh; represented on the left axis of the figure) and 'battery density' (measured in Wh per litre; represented on the right axis) has evolved between 2008 and 2016. The effect of the battery cost on the BV flow is negative and non-linear, while the relationship with the battery density is positive and shows a somewhat more linear behaviour. As the batteries become cheaper and their performance improves, the electric car will gain ground over the traditional car.

The new battery developments, which are currently in the R&D phase, the expansion in battery production volumes (economies of scale) and the growth of their storage capacity should allow a significant drop in the per-unit costs of these storage systems in the coming years. According to the U.S. Department of Energy (DOE), increasing battery production volumes from 25,000 to 100,000 units for 100 kWh BV would reduce battery production costs (per kWh) by 13%. Several studies also suggest this direction, such as those by Howell [7] and Slowik et al. [11]. On the other hand, according to Howell [7], increasing the size of the battery from 60 to 100

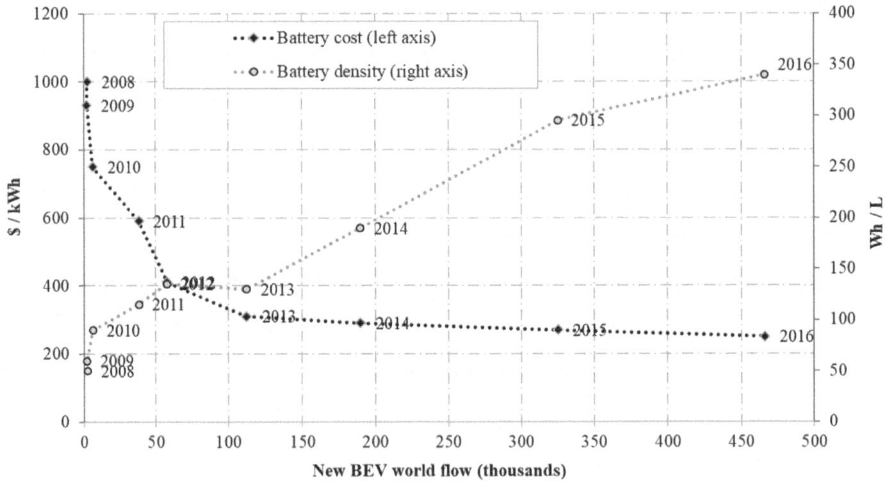

Fig. 5 Relationship between the flow of new VB and the battery cost and density. 2010–2016. *Source* US Department of Energy

kWh (which roughly reflects, for an average car sold in the US, an increase in the road range from 200 to 320 km) would lead to a 17% reduction in the battery cost per kWh.

As regards the second determinant, the development of the auxiliary industry, a good proxy of the behaviour of this industry is given by the evolution of the deployment of charging points, which can be fast or slow and be, as well, in public or private spaces. Figure 6 shows the percentage distribution of fast and slow chargers by country in the years 2010 and 2015.

In 2010, Japan (59.4%), China (23.4%) and, to a lesser extent, the US (11.4%) had the fastest chargers worldwide—in fact, China and Japan only had fast chargers. Slow chargers are mainly (in 2010) Norway (55.8%), Italy (12.2%), the USA (9.6%)

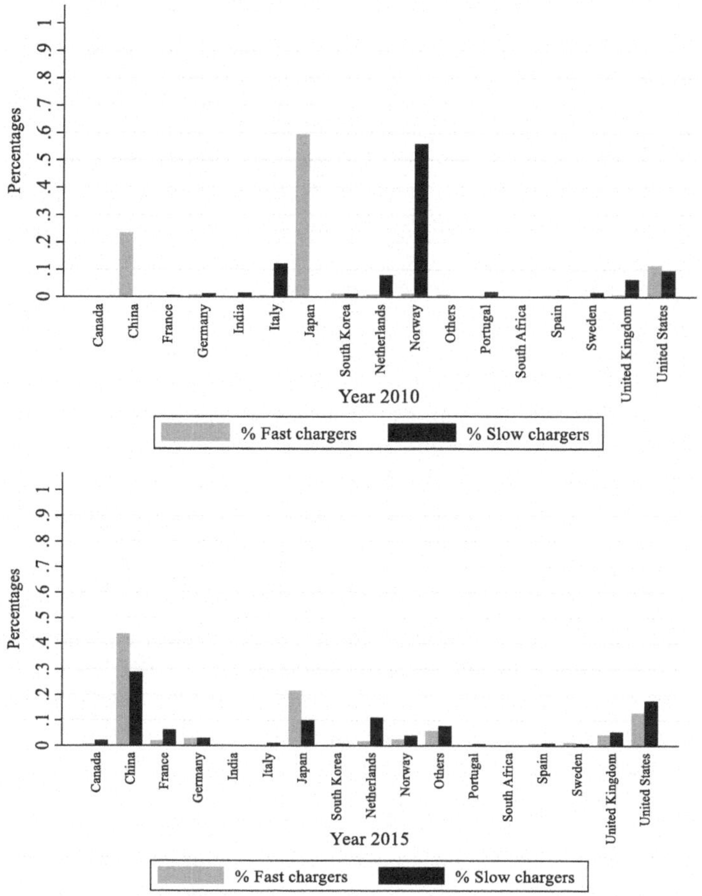

Fig. 6 Evolution of the battery charger by country. 2010 versus 2015. *Source* Own elaboration based on IEA data

and the Netherlands (8%). Unlike the USA and the Eastern countries, most European countries had opted almost exclusively for this type of (slow) charger in 2010.

In the year 2015, the picture looks very different. China, with a 43.7% share, ranks as the country with the highest percentage of fast chargers, to the detriment of Japan (21.6%); the USA increases its share somewhat compared to 2010 (from 11.4% in 2010 to 12.7% in 2015). As for the slow chargers, China (28.8%), the USA (17.4%), the Netherlands (11%) and Japan (10%) are the dominant countries. In Europe and the USA, unlike what happens in Eastern countries, these (slow) chargers have even (in 2015) a greater weight in the world stock than that shown by the fast chargers. Another fact to highlight, considering all the countries together, is that the ratio of slow chargers to each fast one is approximately 5.8 in 2015—this ratio was 9.6 in 2010. Spain is above this aggregate value in 2015, as it has 8 slow chargers for each fast one—the same ratio as countries such as the United States or the United Kingdom.

Another interesting relationship related to charging points is shown in Fig. 7, where the 2010–2015 evolution of the relationship between BV registrations and the number of chargers, both fast (left panel) and slow (right panel)—between new BVs and new HVs, we focus on the new BVs because they depend more on the existence of chargers. The graph shows two important features: (1) new BV registrations are growing continuously in all countries, with the exception of Japan and the Netherlands in 2015, when registrations are falling somewhat from the previous year; (2) the effect of an increase in fast chargers on new registrations of BV appears to be greater than that of slow chargers, as shown by the steeper slopes in the graph relating BV to fast chargers, compared to the graph of BV to slow chargers—the econometric model in the next section will allow us to explore these relationships more precisely.

Figure 7 also allows the estimation of the average slope or 'new BV/fast chargers' ratio in the different years and countries. Thus, in 2010, countries such as Norway, Germany, Spain and the United Kingdom showed a high ratio of new VB per fast charger, a ratio greater than 30 in all three cases. In Norway, for example, there were 65 new registrations of BV per quick charger (in Germany 47 and Spain 35). In the year 2015, the picture changes. In 2015, Portugal, Norway and France are the countries with the highest ratios (all with more than 30 new registrations per quick charger), while in countries such as Spain, the Netherlands, the United Kingdom and Germany the ratio falls significantly with respect to 2010, to below 10 new registrations per quick charger in the case of the first three countries and 15 new registrations per quick charger in the case of Germany.

The third and last determinant factor of the electric car's progress is the existence of policies to encourage the supply and demand of EVs and the auxiliary equipment industry (mainly, charging points). Various public policies are promoting the purchase of EV and the development of publicly accessible charging points, financed through direct investment or public–private initiatives. Some of these initiatives to generate charging points in public spaces go beyond urban areas, giving rise to charging networks that allow long-distance EV trips even on a continental scale; for example, at a European level we can mention initiatives such as NewMotion or Electrek.

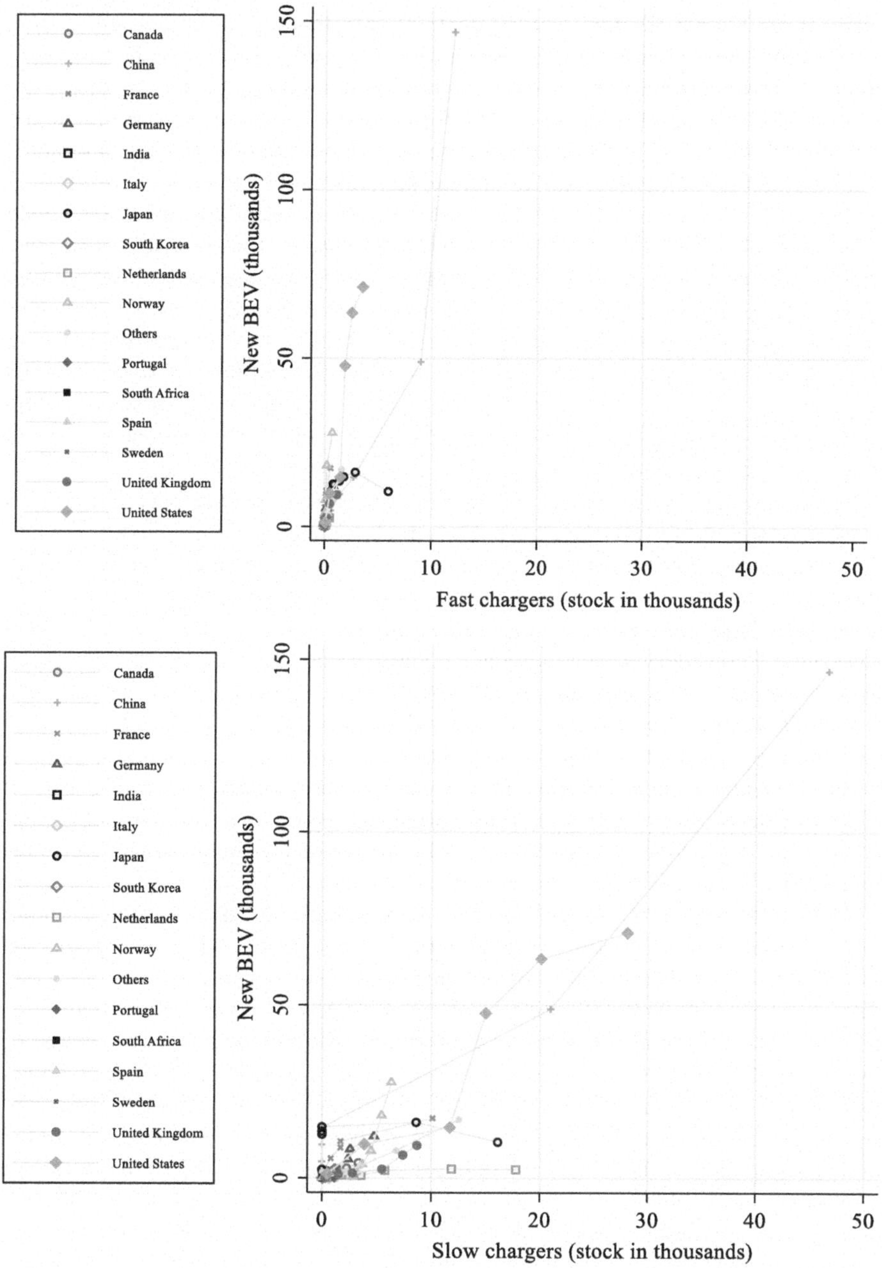

Fig. 7 Ratio of chargers (fast and slow) to new BV by country. Period 2010–2015. *Source* Own elaboration based on IEA data

The IEA's Global EV Outlook [6] contains information on the EV stimulation policies implemented in the different countries analysed in this study. In particular, a distinction is made between 'market creation' policies and policies supporting the 'creation of charging infrastructure'. The former includes measures to stimulate the supply and demand of EV. Those aimed at the supply side includes financial measures (such as direct investment incentives and tax benefits) and regulatory measures (e.g. regulation of tailpipe emissions or fuel saving standards). Similarly, measures aimed at the demand side of EVs may be of a financial nature (e.g. 'renove' operations, subsidized prices or tax credits) or of a regulatory nature (e.g. tax exemptions, parking fee and toll exemptions, and access permits to certain areas and lanes). On the other hand, policies to support the creation of cargo infrastructure basically consist of direct incentives to investment by public authorities and institutions, or tax benefits such as tax breaks for individuals or private entities for the installation of loading points. All these policies, in turn, can be carried out at the local, regional or national level, covering a different percentage of the population of each country.

As an attempt to synthesize the position of each country with respect to two types of financial support referred, we have developed a cluster analysis which takes into consideration the similarities among countries according to the measures they adopt. Specifically, we follow a hierarchical method of successive grouping of countries into increasingly large clusters according to the similarity of the policies they adopt. That is, we start by considering each individual country as a separate group, then we group the two countries with most similar policies, etcetera; so the clustering process can be continued until all countries get integrated into a single group. This hierarchical method allows us to attain a specific number of groups since the clustering process can be interrupted once that number is reached. Furthermore, the results can be shown in a dendrogram, a diagram that outlines the successive formation of greater and increasingly heterogeneous groups—the height of the dendrogram represents dissimilarity.

For comparative purposes, we have drawn up two different clusters using, as input data, information on measures to stimulate the EV in the year 2015 (see Fig. 8). One cluster is based on the similarity of policies of market creation, and the other one is based on the similarity of support policies to create charging infrastructure. In order to ascribe a numerical value to each policy, we have weighed the degree of coverage of the national territory; so, a value 0 has been ascribed when a given policy has not been implemented within a considered country; 0.25 when the policies are implemented within certain geographical regions with less than 50% coverage of the country's population; 0.75 when the policies cover more than 50% of the country's population; and 1 when the policies cover the whole country's population. Table 1 summarizes the measures (and their respective coverages) in the analysed countries.

Figure 8 shows the resulting dendrograms. The cluster based on policies of incentive to demand and supply of electric vehicles (panel on the left of the figure) shows four country clusters at least: (I) Canada, Italy and India; (II) China, Japan, South Korea and Portugal; (III) France, United Kingdom and Netherlands and (IV) Denmark and Norway. The rest of the countries (Spain among them) can be considered units relatively different, since they join other countries (or their clusters) at

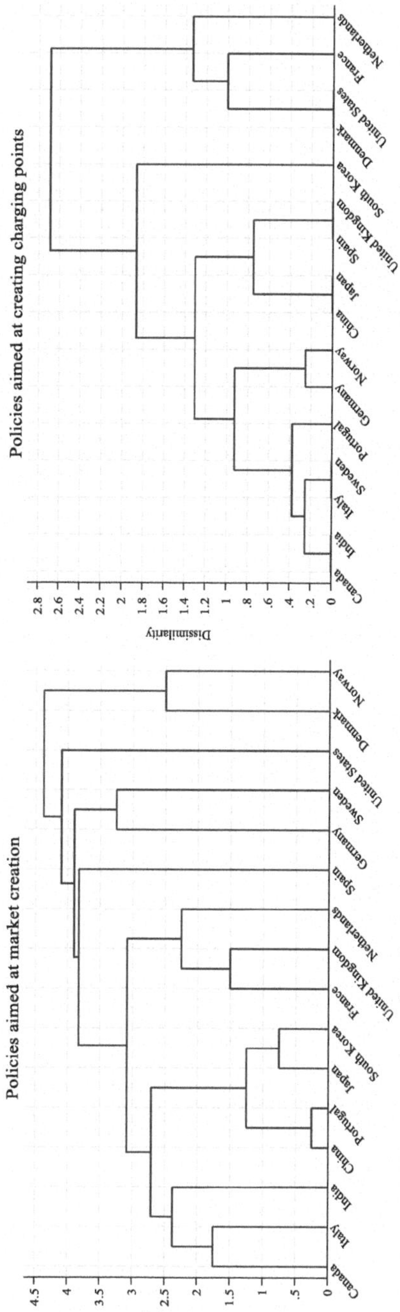

Fig. 8 Clusters of countries according to stimulus measures to the electric car. Year 2015. *Source* Own elaboration based on IEA data

Table 1 International comparison of EV support policies. Year 2015

	Policies aimed at market creation												
	EV purchase incentives				EV use and circulation incentives				Waivers on access restrictions			Tailpipe emissions standards	
Country	Rebates at registration/sale	Sales tax exemptions (excl. VAT)	VAT exemptions	Tax credits	Circulation tax exemptions	Waivers on fees (e.g. tolls, parking, ferries)	Electricity supply reductions /exemptions	Tax credits (company cars)	Access to bus lanes	Access to HOV lanes	Access to restricted traffic zones	Fuel economy standards/regulations including elements	Road vehicles tailpipe pollutant emissions standards
Canada	0.75	0	0	0	0	0.25	0	0	0	0.25	0	1	Tier 2
China	1	1	0	0	1	0.25	0.25	0	0	0	0.25	1	China 5
Denmark	0	1	0	0	0	0.75	0	0	0	0	0	1	Euro 6
France	1	1	0	0	0	0.25	0	1	0.25	0	0.25	1	Euro 6
Germany	0	0	0	0	1	1	0	1	0.25	0	0.25	1	Euro 6
India	1	0.75	0.75	0	0.25	0	0	0	0	0	0	1	Bharat 3
Italy	1	0	0	0	1	0	0	0	0	0	0	1	Euro 6
Japan	1	1	0	0	0.75	0.25	1	0	0	0	0	1	JPN 2009
Netherlands	0	1	0	0	1	0	0	1	0	0	0.25	1	Euro 6
Norway	0	1	1	0	0	1	0.25	0	1	0	0	1	Euro 6
Portugal	1	1	0	0	1	0.25	0.25	0	0	0	0	1	Euro 6
South Korea	1	1	0	0	1	0.75	1	0	0	0	0	1	Kor 3
Spain	1	1	0	0	0.25	1	0	1	0	1	0	1	Euro 6
Sweden	1	0	0	1	1	0.25	0	1	0	0	0	1	Euro 6
United Kingdom	1	1	0	0	1	0.25	0.25	1	0.25	0	0	1	Euro 6
United States	0.25	0.25	0	1	0.25	0.25	0.25	0	0	0.25	0	1	Tier 2

(continued)

Table 1 (continued)

Country	Policies aimed at creating charging points			
	Direct investment		Fiscal advantages	
	Publicly accessible chargers	Private chargers	Publicly accessible chargers	Private chargers
Canada	0.25	0.25	0	0
China	1	1	0	0
Denmark	1	0	1	1
France	1	1	1	1
Germany	1	0	0	0
India	0.25	0.25	0	0
Italy	0.25	0	0	0
Japan	1	1	0	0
Netherlands	1	0	0	1
Norway	1	0.25	0	0
Portugal	0	0	0	0
South Korea	0.25	1	0.25	0.75
Spain	0.25	1	0	0
Sweden	0.25	0	0	0
United Kingdom	0.25	1	0	0
United States	1	0	1	1

elevated heights in the dendrogram. Regarding the cluster based on policies to the creation of chargers (panel on the right), it shows five clusters well differentiated: (I) Canada, India, Italy, Sweden and Portugal; (II) Germany and Norway; (III) China and Japan; (IV) Spain and United Kingdom and (V) Denmark, USA, France and Netherlands—South Korea arises as a relatively different country with respect to the development of this kind of policies. In the following section, we will compare these results to those obtained from a frontier model for panel data which allows us to measure the efficiency of countries in the implementation of BEVs.

3 Stochastic Frontier Panel Data Model. Estimation by Countries

The Electric Vehicles Initiative [3] stipulates as a goal a fleet of 20 million EV worldwide for 2020. On the other hand, the Paris Declaration on Electro-Mobility, Climate Change and Call to Action establishes a global goal of deployment of 100 million EV and 400 million 2 and 3 wheel EV in 2030. The achievement of these goals implies a substantial growth of the market in order to develop much more the current stock of 2 million EV in 2016.

As we have discussed in the previous section, the achievement of the described goals in each country fundamentally depends on economic, technological and political variables. Put another way, the evolution of the new licence plates of this type of vehicles will depend on the evolution of its production costs (as well as its relative price in comparison with the traditional fuel cars), on the parallel evolution of its supplementary industry (repair, maintenance, charging points, etc.) and on the support policies to EV and charging points defined by the governments. The available data in this study will allow us to find evidence about the impact of such factors on the registration of BEVs—as we explained in the previous section, we are focusing on the new BEV and not on the PHEV, because the former depends to a greater extent on the existence of charging points.

To determine the impact of the different variables on the registration of BEV, we estimate both a panel data with fixed effects and a stochastic frontier model for panel data. We consider six years (period 2010–2015) and 15 countries in the panel; countries for which complete data is available, which are: (1) Canada, (2) China, (3) France, (4) Germany, (5) India, (6) Italy, (7) Japan, (8) Korea, (9) Netherland, (10) Norway, (11) Portugal, (12) Spain, (13) Sweden, (14) United Kingdom and (15) US. We consider the flow of new BEV (yearly registrations) as the variable to be explained. Among the explanatory variables, we control for the stocks of both slow and fast chargers, and for the cost and density of the battery—these two variables will vary year to year but not country to country, i.e. they have intra-group variability, but not between-group variability. All variables are expressed in logarithms in order to control possible non-linear relationships among variables; this way, the estimated coefficients represent elasticities.

The panel data is estimated with fixed effects to properly deal with the existence of correlation among the regressors and the individual effects. Likewise, it is worth recalling that in this kind of econometric models the individual effect of each country controls for those (time-constant) features of the country that are not directly observed in the panel (in the data), but that have influence on the dependent variable, like, for example, the guidelines of each government regarding the regulation and incentives to the EV industry.

For its part, like panel data models, stochastic frontier models make possible to estimate the parameters of a linear model of panel data by defining a perturbation that follows a mixed distribution with two components: one that follows a strictly positive distribution (this restriction of non-negativity is the main difference between the stochastic frontier panel data and the conventional panel data), and another one that follows a symmetric random distribution. In econometric literature, the non-negative component is known as the inefficiency term of each sample unit, and the component with the symmetric distribution as the idiosyncratic error corresponding to each sample unit in each sample period. The model allows two different parameterizations of the inefficiency term: time-varying and time-constant. In the model of time-constant inefficiencies the inefficiency term is supposed to have a normal truncated distribution. In the parameterization of the model with time-varying inefficiencies (model of [1]), the inefficiency term is configured as a normal-truncated random variable multiplied by a specific time function (which can be increasing or decreasing). In both models, the idiosyncratic error term is supposed to follow a normal distribution.[2]

In this study, when estimating an output-oriented model of inefficiency, we are assuming that the driving factors of the BEV operate as inputs of a 'production function' whose output is the new yearly registrations of BEV. So the countries that manage to generate a greater flow of registrations with the same input levels are more efficient units. The estimated panel data with time-constant inefficiencies is as follows:

$$\begin{aligned} ln(new\ BEV)_{it} &= ln(\widehat{new\ BEV})_{it} + \hat{v}_{it} = ln(\widehat{new\ BEV})_i^{Frontier} - \hat{u}_i + \hat{v}_{it} \\ &= ln(\widehat{new\ BEV})_{it}^{Frontier} + \hat{\varepsilon}_{it} \\ &= \hat{\beta}_0 + \hat{\beta}_1 ln\ (Slow\ chargers)_{it} + \hat{\beta}_2 ln\ (Fast\ chargers)_{it} + \hat{\beta}_3 ln\ (battery\ costs)_t \\ &\quad + \hat{\beta}_2 ln\ (battery\ density)_t - \hat{u}_i + \hat{v}_{it} \end{aligned}$$

$$\text{where} \varepsilon_{it} = v_{it} - u_i,\ v_{it} \sim N\left(0, \sigma_v^2\right),\ u_i \sim N_+\left(0, \sigma_u^2\right),$$
$$i = 1, \ldots, 15,\ t = 2005, \ldots, 2010$$

Notice that $-\hat{u}_i$ measures how inefficient is sample unit i regarding its frontier $ln(\widehat{new\ BEV})_i^{Frontier}$ once controlled the effect of the random perturbation $\hat{v}_{it}$ that

[2] On panel data methodology, see for example Cameron and Trivedi [2] and Kennedy [8]. On stochastic frontier models for panel data see Kumbhakar et al. [9].

might have affected it each period. We have chosen the model with invariant inefficiency because the time-varying alternative estimates a coefficient for the temporal dynamics of the inefficiency term very close to zero (−0.04), which indicates that the inefficiency of each country does not change much over time; note that the reference interval is not very wide (2010–2015), so it is not very probable to find a relevant technological change within the sector.

Table 2 synthesizes both estimated models, the panel data and the panel data with inefficiency.

The estimated coefficients from both models support similar conclusions. The new registrations of BEV fundamentally depend on the number of fast chargers, and show no dependency with respect to slow chargers. One issue to take into account is that the causality between the variable BEV and the variable 'charger points' can be confusing and might give rise to a problem of endogeneity, since certain omitted variables in the model (e.g. incentive measures to the EV industry) might simultaneously affect both variables. In any case, we think the variable 'charger points' must be considered into the model just as one more regressor, since it seems clear that investors (both public and private) have recognized the implementation of a sufficient network of charging points as a driver behind the demand of BEV.

If we compare both models, it becomes appreciable that the elasticity of the BEV in relation to the fast charging points is greater in the model with efficiency (0.28 versus 0.15). A coefficient close to 0.3 would indicate that a 10% increase of fast charging points could increase BEV registrations by about 3%.

The other determining factor of the deployment of the BEV is the battery cost. According to our estimations, the relation between the BEV new matriculations and the battery cost is negative and elastic, so a 1% decrease of the cost would allow to increase the volume of new BEV more than 1% (between 1.3 and 1.7%).

One of the advantages of the panel data model with inefficiency, in contrast to the panel data with fixed effects, is that it allows to estimate the global efficient frontier (the 'production function' of BEV) and the distance of each country to that frontier, once removed the effect of possible random perturbations that fall outside the control of the country. Figure 9 shows the position of the different countries in relation to the overall efficient frontier, both in terms of the actual series of new BEV (in logarithms) and in terms of the model prediction of that variable, which removes the term of random perturbation v_{it} from the actual series. The figure shows that US, France, China, Japan and Norway are the most efficient countries (closest to their efficient frontier), whereas South Korea, Spain, Italy and Portugal are the most inefficient ones.

If we compare the individual effects of the panel data with fixed effects to the inefficiencies of the panel data with frontier (Fig. 10), an expectable but interesting result arises: those countries that show a lesser degree of inefficiency in the panel data with frontier show as well a greater individual effect in the panel data with fixed effects. This result is expectable if we consider that when a given country reveals a high individual effect, the expected value of the endogenous variable in the fixed effects model has a constant term for that country which is greater than the overall constant term of the model.

Table 2 Estimates of the determinants of BEV registrations. Period 2010–2015

Model with time-constant inefficiency. Variable explained: BEV registrations (in logarithms)

	Coef.	Std. Err.	z	P > z	[95% Conf. Interval]	
Log (slow chargers)	0.072	0.11	0.67	0.50	−0.14	0.29
Log (fast chargers)	0.28***	0.09	2.97	0.00	0.09	0.46
Log (battery cost)	−1.32***	0.41	−3.22	0.00	−2.12	−0.52
Log (battery density)	0.370	0.36	1.03	0.30	−0.33	1.07
Constant	9.04**	3.86	2.34	0.02	1.47	16.62
μ	1.56**	0.66	2.36	0.02	0.27	2.85
$\log(\sigma^{2)}$	0.38	0.52	0.74	0.46	−0.64	1.40
Inverse of the logit of γ	1.7**	0.66	2.56	0.01	0.40	3.00
σ^2	1.47	0.76			0.53	4.07
γ	0.85	0.09			0.60	0.95
σ_u^2	1.24	0.77			−0.26	2.74
σ_v^2	0.23	0.04			0.14	0.31
Number of observations	76					
Number of countries	15					
Wald chi2(4)	397.31 (Prob > chi2 = 0)					
Log likelihood	−74.0					

Model with fixed-effects. Variable explained: BEV registrations (in logarithms)

	Coef.	Std. Err.	t	P > t	[95% Conf. Interval]	
Log (slow chargers)	0.047	0.14	0.32	0.75	−0.26	0.36
Log (fast chargers)	0.15*	0.10	1.54	0.1	−0.06	0.37
Log (battery cost)	−1.69**	0.58	−2.95	0.01	−2.93	−0.46
Log (battery density)	0.53*	0.33	1.59	0.1	−0.18	1.25
Constant	8.32*	4.46	1.87	0.08	−1.25	17.90
σ_u	1.21					
σ_ε	0.48					
Coefficient	0.86 (fraction of variance due to u_i)					
R^2 within	0.864					
R^2 between	0.655					
R^2 overall	0.584					
Number of observations	76					
Number of countries	15					
F(4,14)	80.66 (Prob > F = 0)					
corr(u_i, Xb)	0.2183					

*p<0.1; **p<0.5; ***p<0.01

Source Own elaboration based on IEA data

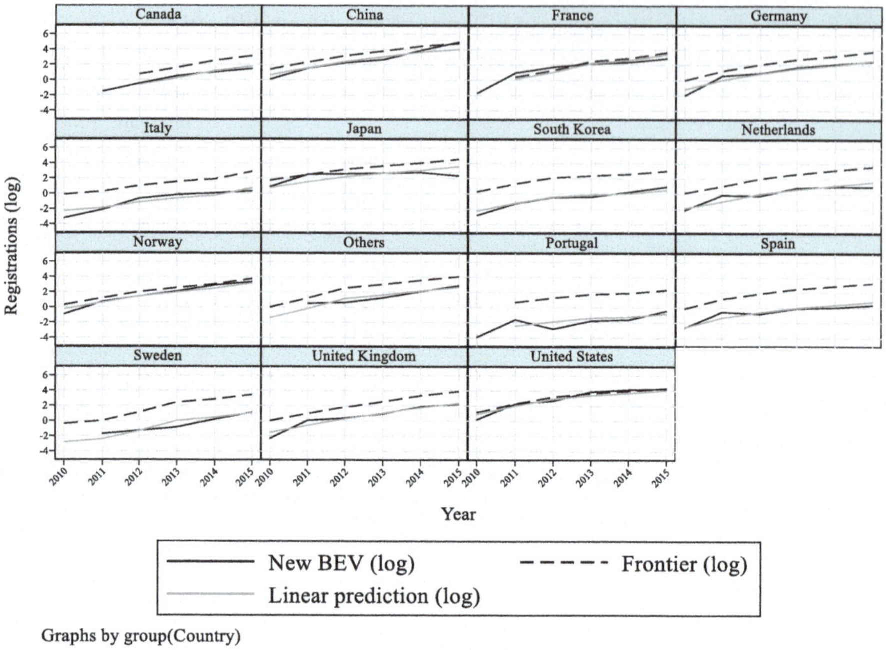

Fig. 9 Positioning of countries regarding the overall efficient frontier. *Source* Own elaboration based on IEA data

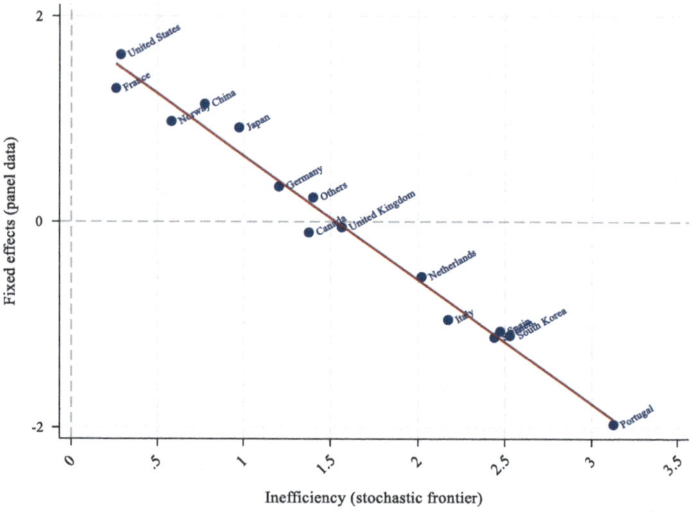

Fig. 10 Inefficiencies against fixed effects. *Source* Own elaboration based on IEA data

We conclude this analysis on the deployment of the BEV searching for some evidence about the effectiveness of the support policies for the EV. To that end, we have compared the estimated individual effects and efficiencies from the panel data models (see Fig. 10) to the two clusters generated by similarity of support policies for the EV. In such comparison, we observe a greater similarity between the results of the panel data models and the cluster based on policies aimed at the creation of charging infrastructure (vs. the cluster based on policies of market creation). As it turns out, the six most efficient countries in the panel data models are related to three clusters relatively homogeneous: USA and France (together with Denmark); China and Japan; and Germany and Norway. This result allows us to derive at least two conclusions. First, the efficiency of the deployment of the BEV seems to be more related to the policies of creation of supplementary industry (charging points) than to those of incentives to the market. Second, the policies of the most efficient countries (or country clusters) should be a valid reference to other countries in order to achieve a greater efficiency in the deployment process of the electric vehicle.

4 Conclusions and Industrial Policy Recommendations

Important differences among countries are observed in the introduction of the electric vehicle. We can highlight the case of Norway, where almost 40% of new registrations are related to this technology, in contrast with countries like Spain, Italy or Greece, where this percentage is less than 0.5%.

In order to find out the factors that determine the number of registrations of battery electric vehicles, we have developed a panel data analysis that points out the cost of the battery and the existence of a public network of fast chargers as the most relevant explanatory variables. Regarding the cost of the battery, we can consider it a common factor to all the analysed countries, and it manifests an elastic condition, with elasticity values of roughly -1.5, that is, for every 10% decrease of the battery cost the number of registrations will increase up to 15%, which makes sense since it is the most relevant cost in the production of the battery electric vehicle, and the consumers manifest a high sensitivity to the initial payment for the vehicle acquisition.

Regarding the existence of a network of fast charging points, it could be determined by the existence of a suitable policy of public promotion, which would be reflected in each country's individual effects (in our econometric proposal). In any case, it has been identified by the econometric analysis as a relevant factor, very significant for the deployment of the electric vehicle, which shows an elasticity close to 0.3%, which means that an increase of 10% of the fast charging points could increase battery electric car registrations almost 3%.

The countries with the greatest individual effects in the estimated models are USA, France and Norway, all of which implement incentives to the purchase and use of the electric vehicle and give a clear support to the development of a fast charging network of public use.

References

1. Battese GE, Coelli TJ (1992) Frontier production functions, technical efficiency and panel data: with application to paddy farmers in India. J Product Anal 3:153–169
2. Cameron AC, Trivedi PK (2009) Microeconometrics: methods and applications. Cambridge University Press, Cambridge
3. Electric Vehicles Initiative, EVI-20-20
4. Global EV Outlook (2018): Towards cross-modal electrification. In: International energy agency, OCDE, París, Francia, pp 1–140
5. Global EV Outlook (2017) Two million and counting. In: International energy agency, OCDE, París, Francia, pp 1–66
6. Global EV Outlook (2016) Beyond one million electric cars. In: International energy agency, OCDE, París, Francia, pp 1–39
7. Howell D (2017) Personal communications of the authors on the status and projections for EV batteries RD&D
8. Kennedy P (2008) A guide to Econometrics, 6th edn. Wiley-Blackwell, Massachusetts, USA
9. Kumbhakar SC, Lovell CAK (2000) Stochastic frontier analysis. Cambridge University Press, Cambridge
10. NewMotion, Europe's largest electric charging partner. https://newmotion.com
11. Slowik P, Pavlenko N, Lutsey N (2016) Assessment of next-generation electric vehicle technologies. In: White paper, International Council on Clean Transportation, pp 1–18
12. UBS Q-Series (2017) Evidence lab electric car teardown—Disruption ahead?

Economic Analysis of Recharging Electric Vehicles

Angel Arcos-Vargas and Antonio Hidalgo

Abstract This chapter analyzes the business model of electric vehicle recharging. As seen in previous chapters, the existence of a fast charging network is a success factor for the penetration of electric vehicles. Although, under current conditions, it is difficult to justify the provision of this service from a single-product private company point of view. The profit and loss accounts of these potential companies are considered, including possible alternatives to improve their results. Another possibility could be that the business model is based on the cross-selling of other products (restaurants, department stores, ...) for which another type of more complex analysis would be necessary.

1 Introduction

Road transport is a key sector for the decarbonization of any national economy but requires that internal combustion vehicles be replaced with electric vehicles (EV). Although there is no need to hide the difficulty of implementing this substitution, several studies have analyzed the impact of this substitution from different technological [8], economic [1], environmental [2], and employment perspectives [3].

With the current technology, the consumption of an electric vehicle is between 12 and 20 kWh per 100 km in a context of batteries with a capacity of between 15 and 40 kWh. With such ranges, an electric vehicle enjoys an autonomy between 100 and 300 km, which differs mainly by the type of driving. Therefore, the circulation of electric vehicles (with the exception of hybrids) requires the execution of a wide network of recharging points; consequently, the main variables in this type of vehicles are related to the operation of the batteries, which can be recharged in different ways depending on their specifications.

A. Arcos-Vargas (✉)
University of Seville, Seville, Spain
e-mail: aarcos@us.es

A. Hidalgo
Universidad Politécnica de Madrid, Madrid, Spain

A. Arcos-Vargas (ed.), *The Role of the Electric Vehicle in the Energy Transition*,
Green Energy and Technology, https://doi.org/10.1007/978-3-030-50633-9_5

In the market for electric vehicle charging infrastructures, there are currently three types of recharging of the batteries depending on the recharging speed, which is determined by the type of electric current (alternate current single-phase or three-phase, or direct current):

- Slow recharge—modes 1 and 2, with a duration between 8 and 10 h (230 V single-phase alternating current and 16 A intensity).
- Fast recharge—mode 3, with a duration between 1 and 6 h (400 V three-phase alternating current and 32 A intensity).
- Ultrafast recharge—mode 4, with a duration of 30 min (400 V of direct current and 43 kW of power).

In general, the single-phase-three-phase 230–400 V AC power sockets, with an intensity of 16–32 A and 2–11 kW of power, are the most common, and allow battery charging that can last several hours depending on the combination of these variables. For its part, the installation of a DC outlet, generally associated with high voltages, intensities, and powers, allows charging in less than an hour, but requires a specific infrastructure and supervision, usually linked to commercial exploitation or use-intensive electricity, such service stations aimed at recharging electric vehicles [6].

The objective of this chapter is to study the economic viability of the installation of a charging post in Spain. For this, the significant variables of the model have been defined according to empirical, statistical, and regulatory data and only electric battery vehicles have been considered. The hypothesis adopted in the case of hybrid vehicles is that they perform recharging at their home or company (private recharging), using the internal combustion engine whenever they need to extend their autonomy on public roads.

The rest of the chapter is structured as follows: Sect. 2 identifies the hypotheses that serve as the basis for the study; Sects. 3 and 4 analyze estimates of income and operating costs, respectively, the simulation of results is conducted in Section, and the main conclusions are laid out in Sect. 6.

2 Main Hypotheses

For the purpose of estimating the economic analysis of the recharging of an electric vehicle, a model of a fast alternating current recharge post for three-phase and external mounting networks was chosen for the calculations. The nominal recharge power is 22 kW, in mode 3, with two Mennekes type 2 connectors for simultaneous use. These connectors, which are approved as European standard, allow single-phase loads up to 16 A and three-phase loads up to 63 A, resulting in power of 3.5 kW and 44 kW, respectively.

Although there are numerous alternatives on the market, the RVE-PT3 v10425 Circutor model was adopted for the purpose of simulation. This post model is suitable for outdoor locations, such as public roads, and public or large outdoor parking lots, airports, etc., In addition to electrical safety systems, this post has specific characteristics in terms of robustness, both variable environmental conditions and vandalism, being compatible with all types of vehicles and recharge modes 1 and 3.

While the life of a recharge post depends on the model and manufacturer, all posts on the market generally have a recommended service life of between seven and 10 years. In this study, the most conservative case (that is, a useful life of 10 years) has been considered for the hypothesis, although other studies have considered more unfavorable scenarios, such as 7.5 years [4].

Including the cost of the equipment and its installation, which incorporates all the work and the necessary auxiliary electrical equipment, the price rises to approximately €30,000, not counting the extension of the network. This post model has been chosen because it is one of the simplest solutions that meet the public recharge needs of users. Considering the installation of equipment compatible with the four modes of recharging in direct current and with similar characteristics, the investment would amount to €50,000, also increasing the other costs incurred.

Once the location of the charging post has been requested, it is the distribution company that is responsible for defining the connection point to the network and for carrying out the extension from that point to the location. Assuming that the power request will be less than 100 kW, the cost of the extension is the product of the price of the scale of the extension (€28 per kW), estimated as a representative value of the associated average costs, by the amount of kW. When considering a nominal recharge power of 22 kW, the cost of the extension will amount to €616. However, in the event that the location point is in an interurban area (belt, highway, road, etc.), these costs may be substantially higher.

3 Income Estimate

The energy supplied by a post depends on the number of vehicles that use it, the frequency with which they perform a recharge, and the unit consumption of each car model. To carry out the estimation of this energy, the data included in Table 1 are used.

Table 1 Initial data related to consumption

Number of vehicles per post (1)	10
Average daily distance traveled (km) (2)	39
Consumption per km (kWh) (3)	0,2
Medium load (kWh) (2)	10
Public recharge (%) (2)	25

Source (1) Directive on the deployment of alternative fuels infrastructure (AFID), focus on electromobility. Directiva 94/2014/UE; (2) Amount of energy in kWh that users make on average in a public post. Zem2all Final report 2016; (3) Markkula et al. [5]

The energy consumed by each vehicle (E_{EV}) is calculated as the average daily distance traveled (D_{Dia}) per unit consumption (E_{km}) for a whole year:

$$E_{EV} = D_{Dia} \cdot E_{km} \cdot 365 \rightarrow E_{EV} = 2.847 \text{ kWh}$$

The average energy supplied by a pole for one year (E_{PR}) is obtained by the product of the number of electric vehicles in circulation per recharging point (EV/PR), the average energy consumed per vehicle, and the utilization rate of a point of public recharge (RP):

$$E_{PR} = \left({}^{EV}/_{PR}\right) \cdot E_{EV} \cdot \text{RP}\ (\%) \rightarrow E_{PR} = 7.118 \text{ kWh}$$

For simplicity, the variability of electricity prices during the day has not been considered. For later studies, the points, valleys, and plains of temporary price fluctuations should be considered, since they constitute an incentive that is provided for in Spanish legislation (RD 647/2011).

As the experience in the electric vehicle recharge market is reduced, it has been hypothesized that the variable cost per kilometer of the electric vehicle, charging the batteries in public recharge posts, must not exceed the lower equivalent cost of traditional fuels, since being superior would not incentivize the penetration of the electric vehicle.

Table 2 shows the characteristic values of price per kilometer (including VAT and special taxes) using diesel and gasoline as fuel, taking into account the characteristic consumption of the current vehicle fleet in Spain, which is 12 years old and mixed road-urban consumption.

Taking into account that a kilometer of diesel fuel is the most economical, the price considered in the base scenario to be passed on to the final customer per kWh would be:

$$P_{kWh} = {}^{P_D}/_{km} \cdot C_{km} \rightarrow P_{kWh} = 0{,}40\ €$$

where C_{km} represents the average electrical consumption per kilometer of an electric vehicle, which is 0.2 kWh/km.

Because the prices of the products do not remain the same throughout the years, annual inflation of 1% has been assumed. This value has been adopted so that the price of the recharge for the final consumer in real terms remains constant throughout

Table 2 Consumption and prices of petroleum derived fuels

Fuel	Consumption/100 km (l)	Price per liter (€)	Cost/km (€)
Diesel	7	1,1	0,08
Gasoline	9	1,3	0,14

Source Monzón et al. [7]

the analysis period. However, the prices of fuels and electricity do not evolve with inflation, but with the evolution of regulation and with the variations in supply and demand in wholesale markets, which can alter the results.

4 Estimation of Operating Costs

4.1 Variable Costs—Energy Purchase

As with income, the price of electricity is not the same throughout the day as there is temporary discrimination at peak and valley hours, and it also depends on the contract with the trading company. However, this analysis does not discriminate temporarily or at the time of recharge, or in the price of electricity.

Since the charging post is fed from the low voltage network and the contracted power is greater than 15 kW, there is no regulated rate that can be applied, and must go to the free market. In this context, the price of energy (variable cost) will have three components: access fee (3.0A), energy (including the marketer's margin), and taxes (VAT and special electricity tax). Table 3 shows, for illustrative purposes, the offer of a marketer (Endesa Energy) that is adapted as a reference for the analysis, considering that the consumption is carried out in a homogeneous way throughout the day.

If each post supplies an average of 7118 kWh, the variable costs in the first year are:

7.118 kWh x 0,149983 €/kWh = 1.067,58 €

Table 3 Endesa Energy's offer for low voltage supplies > 15 kW

Product name		System tariff (A)	Endesa (supplier) proposal	
			Peak price (€/kW month)	Energy price (€/kWh)
Power > 15 kW	Increasing savings choice	3.0	6.658935	Peak 0.180649
				Flat 0.145598
				Valley 0.102708
	Tariff choice	3.0	6.672253	0.149983

Table 4 Fixed costs

Posts per operator	30
Annual cost per operator	30.000 €
Maintenance per year	1% of the investment
Communications	60 €
Power quota per kW and month	6,67 €
General expenses	10%

4.2 *Fixed Costs*

The associated fixed costs include the payment of the power quota, the salaries of the staff, the cost of the communication systems, the cost of maintenance, and other general expenses. Table 4 shows the assumed costs for this analysis.

(a) **Power quota (Fix electricity cost)**

The power quota is one of the most important costs because the analysis case is a high-performance receiver. The cost of the quota will be the product of the scale of the quota ($b_{CP} = €6.67$) per kW and month. Each year, and for a 22 kW post model, it would be:

$$C_{CP} = b_{CP} \cdot 22\,\text{kW} \cdot 12\,\text{months} \rightarrow C_{CP} = 1.760{,}48\,€$$

(b) **Human resources**

Although recharging is done automatically without the need for specialized personnel, maintenance checks require qualified personnel. Because these posts constitute an important investment installed on public roads, they are not only subject to the dangers of a possible malfunction, but also to the failures caused by the interaction with the environment or any possible case of vandalism. Also, this infrastructure works with high power, so a failure in protection systems, insulation, or overheating can be a potential danger for a person, who is recharging his/her vehicle at that time, or for the environment.

In addition to these potential hazards, it should be considered that insufficient maintenance could result in a breakdown remaining unsolved for a prolonged period. When deciding whether or not to acquire an electric vehicle, the average user may discount the fact that the load posts do not have high reliability (high degree of availability).

According to the information provided by management companies of recharge posts, assuming a cost per company per operator of €30,000 year, and that each operator maintains an average of 30 recharge posts, the cost of personnel for each post is estimated at €1000 per year.

(c) **Communications**

To guarantee correct operation, as well as the interoperability of services between the different actors that take part in the recharging process of an electric vehicle, standardized communication and information protocols that connect the charging posts with the control center are required. The estimated cost of the necessary communications is estimated at €60 per year.

(d) **Maintenance**

As noted above, it is necessary to carry out periodic maintenance work resulting from small breaks, adaptations, and repairs to ensure the correct operation of the refill post. This cost is estimated, as in other industrial processes, at 1% per year of fixed assets; that is €300 per year.

(e) **General expenses and industrial benefit**

Finally, encompassing the entire series of costs that have not been specified in the previous points, an estimated 10% of the sum of all has been estimated. In the case analyzed, this represents a total of €311 per year.

5 Results Simulation

As can be seen in Table 5, according to the assumptions considered in the case of analysis, the flows generated in each year are negative, which implies that the investment is not amortized. Having assumed a price for the sale of energy in the charging posts of €0.40 per kWh, and given that the density of vehicles per post follows the recommendation of the European Directive, two possible options are proposed to ensure a reasonable profitability of the project: (a) increase the sale price of energy, or (b) reduce fixed costs through subsidies or exemptions, taking into account that the most relevant items of fixed costs are associated with the power quota and maintenance personnel expenses.

Each of the possible actions aimed at guaranteeing the profitability of this investment is analyzed below:

- *Minimum allowable price scenario.* Starting from the case analyzed, it is possible to ask what the sale price of the energy in the recharge post should be, assuming that there is no variation in the quantity demanded (rigid behavior; that is, zero demand-price elasticity), so that the entrepreneur can achieve a reasonable return for the assets used (6.5%). The results obtained give a price of €1.21 per kWh, which would be equivalent to a variable cost of €0.24 per km, which is almost three times higher than diesel fuel.

Table 5 Economic analysis simulation (€)

		Year										
		0	1	2	3	4	5	6	7	8	9	10
Energy consumption per vehicle (kWh)			2,847.00	2,847.00	2,847.00	2,847.00	2,847.00	2,847.00	2,847.00	2,847.00	2,847.00	2,847.00
Energy supplied per charging point (kWh)			7,117.50	7,117.50	7,117.50	7,117.50	7,117.50	7,117.50	7,117.50	7,117.50	7,117.50	7,117.50
Incomes			**2,847.00**	**2,875.47**	**2,904.22**	**2,933.27**	**2,962.60**	**2,992.23**	**3,022.15**	**3,052.37**	**3,082.89**	**3,113.72**
Investment		**30,616**										
	Equipment	**30,000**										
	Network development	**616**										
Fix cost			**3,432.97**	**3,467.30**	**3,501.97**	**3,536.99**	**3,572.36**	**3,608.08**	**3,644.16**	**3,680.61**	**3,717.41**	**3,754.59**
	Power charge		1,760.88	1,778.49	1,796.27	1,814.24	1,832.38	1,850.70	1,869.21	1,887.90	1,90.78	1,925.85
	Labour cost		1,000.00	1,010.00	1,020.10	1,030.30	1,040.60	1,051.01	1,061.52	1,072.14	1,082.86	1,093.69
	Telecom		60.00	60.60	61.21	61.82	62.44	63.06	63.69	64.33	64.97	65.62
	Mainteance		300.00	303.00	306.03	309.09	312.18	315.30	318.46	321.64	324.86	328.11
	General expenses		312.09	315.21	318.36	321.54	324.76	328.01	331.29	334.60	337.95	341.33
Variable cost				**1,078.18**	**1,088.96**	**1,099.85**	**1,110.85**	**1,121.96**	**1,133.18**	**1,144.51**	**1,155.95**	**1,167.51**
	Peak energy		352.28	355.80	359.36	362.95	366.58	370.25	373.95	377.69	381.46	385.28
	Valley energy		352.28	355.80	359.36	362.95	366.58	370.25	373.95	377.69	381.46	385.28
	Flat energy		362.95	366.58	370.25	373.95	377.69	381.47	385.28	389.13	393.02	396.95

(continued)

Table 5 (continued)

	Year										
	0	1	2	3	4	5	6	7	8	9	10
Profit and loss account											
Incomes		2,847.00	2,875.47	2,904.22	2,933.27	2,962.60	2,992.23	3,022.15	3,052.37	3,082.89	3,113.72
Fix costs		3,432.97	3,467.30	3,501.97	3,536.99	3,572.36	3,608.08	3,644.16	3,680.61	3,717.41	3,754.59
Variable costs			1,078.18	1,088.96	1,099.85	1,110.85	1,121.96	1,133.18	1,144.51	1,155.95	1,167.51
Depreciations		3,061.60	3,061.60	3,061.60	3,061.60	3,061.60	3,061.60	3,061.60	3,061.60	3,061.60	3,061.60
Gross profit		**−4,715.07**	**−4,731.61**	**−4,748.31**	**−4,765.17**	**−4,782.21**	**−4,799.42**	**−4,816.79**	**−4,834.35**	**−4,852.07**	**−4,869.98**
Taxes		–	–	–	-	–	–	–	–	–	–
Net profit		**−4,715.07**	**−4,731.61**	**−4,748.31**	**−4,765.17**	**−4,782.21**	**−4,799.42**	**−4,816.79**	**−4,834.35**	**−4,852.07**	**−4,869.98**
Cash flow	**−30,616**	**−1,653.47**	**−1,670.01**	**−1,686.71**	**−1,703.57**	**−1,720.61**	**−1,737.82**	**−1,755.19**	**−1,772.75**	**−1,790.47**	**−1,808.38**

- *Scenario of compensation of the power quota. Since the charging* posts are high-power devices, the cost of the fixed power term is significant and determines the size decision. In this way, and in order to favor the deployment of electric mobility, this scenario contemplates the possibility that the regulator, or other agency, exempts said quota, at least temporarily. In this way, fixed costs would be reduced by €1761 per year, generating positive flows, although not sufficient to recover the investment.
- *Scenario for compensation of personnel costs.* This case is symmetrical to the previous one, since it considers that personnel expenses are assumed by another entity, such as the municipality's lighting services. In this way, fixed costs are reduced by €1000 per year, improving the results of the case analyzed, but not enough to generate positive flows.
- *Simultaneous compensation scenario for the power quota and personnel costs.* As in the previous cases, in which the proposed measures are not sufficient individually, this scenario analyzes the effect that both measures would have to apply in aggregate form, which would mean an annual reduction in fixed costs of €2761. Although the system converges on this occasion, it presents negative returns and recovery times of the order of twice its useful life.
- *Simultaneous compensation scenario for power quota and personnel costs plus subsidy.* Starting from the previous case, the percentage of investment to be subsidized is calculated to guarantee the employer a return of 6.5% on the assets. In this case, it is determined that a 67% subsidy is necessary, in addition to the exemption of the power quota and the compensation of the personnel expenses, to guarantee sufficient profitability to the employer.
- *Simultaneous compensation scenario for the power quota and personnel costs, with price increases until the investment is profitable.* Another way to make the recharge business profitable without subsidizing the investment could be to increase the sale price of energy, as was done in the first case, but taking into account that part of the fixed costs are reduced (power share and hand of maintenance work). In the price simulation, it is concluded that, in addition to eliminating the share of power and not supporting labor costs, it would be necessary to sell energy at €0.78 per kWh, which represents almost double the variable cost of diesel vehicles.

6 Conclusions

It is logical to assume that the low presence and slow implementation of the charging infrastructures constitute an important obstacle to the growth of sales of electric vehicles and, in general, for the electrification of road transport that provides many advantages in the horizon of the decrease in emissions to the environment. However, it is difficult to imagine that developed societies willingly admit reductions in consumer comfort levels to achieve these goals. On the other hand, and from an economic perspective, it does not seem foreseeable that users will be able to assume strong increase in the price of products and services, or that the industry will be able to drastically reduce its benefits, which increases the magnitude of the challenge.

In this context, the achievement of a minimum return on the investments made in recharging facilities is considered necessary for the volume of the electric vehicle fleet to accelerate. However, the hypotheses presented in this study present a scenario in which the development of a public network of electric vehicle charging posts obtains a profitability that can be described as doubtful.

In the absence of a number of important electric vehicles that allow a volume of recharges per relevant recharging point and regulatory changes that incorporate subsidies and exemptions to some operating costs, these infrastructures will only make sense if they are used to promote other products or for cross-selling, such as leisure centers, shopping centers, and restaurants.

In our study, and assuming no external help, the price of the energy supplied in the public recharge should be €1.21 per kWh, if a reasonable return for the entrepreneur is pursued, with recharges per point of approximately 7000 kWh per year, which is considered a conservative figure. This value can be 20 times greater than the variable cost of recharging at home at the Supervalle rate and would cost almost a kilometer more than the equivalent of the variable cost of a diesel vehicle.

Another aspect of interest is manifested by the fact that the main components of the cost of these infrastructures are the share of electrical power and the labor required for its maintenance. It is clear that, even when considering a company that operates a greater number of recharging points, it would not modify the results obtained (non-existence of scale effects), since production factors, particularly labor, are divisible, the price of the contracted power has a fixed equivalent value in practice to be regulated, and the bargaining power of the purchase of electric power compared to the trading companies is very small.

In addition, the effect that would increase the penetration of the electric vehicle in the results obtained is not clear. If it were accompanied by a proportional increase in the number of public charging stations, the results obtained would be very similar, whereas, if this increase were less than proportional, the economic results would improve as long as they were associated with a higher level of recharge equipment utilization.

Finally, even assuming that the recharging companies were exempt from the quota of power and maintenance labor, their viability in the case of selling electric power at a price of €0.4 per kWh would require an investment subsidy of 67%, or selling energy at €0.78 per kWh, which would double the variable costs of traditional vehicles, even though the electric vehicle offers other benefits and advantages.

References

1. Granovskii M, Dincer I, Rosen MA (2006) Economic and environmental comparison of conventional, hybrid, electric and hydrogen fuel cell vehicles. J Power Sources 159(2):1186–1193

2. Hawkins TR, Singh B, Majeau-Bettez G, Hammer A (2013) Comparative environmental life cycle assessment of conventional and electric vehicles. J Ind Ecol 17(1):53–64
3. Lindström M, Heimer T (2017) Electric vehicles: shifting gear or changing direction? In: Working paper of the European Foundation for the improvement of living and working condition. FOME Project. Eurofound-European Commission
4. Madina C, Zamora I, Zabala E (2016) Methodology for assessing electric vehicle charging infrastructure business models. Energy Policy 89:284–293
5. Markkula J, Rautiainen A, Järventausta P (2013) The business case of electric vehicle quick charging-no more chicken-and-egg problem. In: EVS27, 27th international electric vehicle symposium & exhibition, Barcelona, Spain, pp 1–4
6. Miranda JM, Iglesias N (2015) Las infraestructuras de recarga y el despegue del vehículo eléctrico. Observatorio Medioambiental 18:57–85
7. Monzón A, Pérez P, Di Ciommo F (2009) La eficiencia energética y ambiental de los modos de transporte de España. TRANSYT, Universidad Politécnica de Madrid, Spain
8. Morais H, Sousa T, Vale Z, Faria P (2014) Evaluation of the electric vehicle impact in the power demand curve in a smart grid environment. Energy Conversat Manag 82:268–282

Environmental Aspects of the Electric Vehicle

Pablo Frías Marín and Carlos De Miguel Perales

Abstract The purpose of this chapter is to understand the complete environmental impact of Electric Vehicles (EVs) compared with traditional combustion engines (ICE). In addition, a review of how European rules have been adapting to cover these impacts is presented. Although the typical approach to the problem is based on the evaluation of the emissions during the use of vehicles, the approach presented here covers the whole life of the Vehicle (manufacturing, use of the vehicle and end of life and recycling) and estimates the amount of material and energy used, the emissions or the toxicity. Results show that EVs have an environmental impact, which is concentrated in the manufacturing phase. Compared with traditional ICEs, EVs have clearly lower emissions when driving, which is certainly critical when defining air quality policies in urban regions. The importance of coordination in environmental policies regional and worldwide is therefore required to guarantee a sustainable and fair transition to a decarbonized transportation.

Currently there is a debate on the environmental impact of Electric Vehicles (EV), based on contradicting information. While the new industry uses the message that EVs are 'Zero Emissions', critics argue that EVs are even more polluting than ICE cars since the Li-Ion battery is not recyclable. To clarify this discussion, and above all give the reader tools for a complete and grounded analysis, this chapter presents different methodologies to assess the environmental impact of EVs, reviews results of the ecological footprint of the EVs during its life and presents the different existing regulations in Europe. By the way, we anticipate that the EV is less contaminated than the traditional combustion engine (ICE).

P. F. Marín (✉) · C. De Miguel Perales
Universidad Pontificia Comillas, Madrid, Spain
e-mail: Pablo.Frias@iit.comillas.edu

A. Arcos-Vargas (ed.), *The Role of the Electric Vehicle in the Energy Transition*, Green Energy and Technology, https://doi.org/10.1007/978-3-030-50633-9_6

1 Evaluation of the Environmental Impact of Electric Vehicles

The recent history of transportation has relied on using fossil fuels (ICE), with different derivatives such as gasoline, diesel and natural gas (liquified or gasified). The combustion of these fossil fuels releases thermal energy, which is converted into mechanical power, and additionally produces exhaust gases released to the atmosphere (carbon dioxide CO_2 and monoxides CO, and nitrogen oxides NO_x, among others). The impressive evolution of road transportation in the last century due to the increasing numbers of trips per person and travelled distances, and the concentration of people in big cities, resulted in high concentration levels of these gases, and generated a real problem of polluted air in major cities in the World [1].

Alternative vehicle technologies based on electricity (battery electric vehicles BEV, hybrid vehicles HEV, or fuel cell vehicles FCEV) are a real solution to the pollution challenge. However, the process of developing these new technologies is not exempt from having an environmental impact throughout its life cycle, from the construction of the different components, its use as a means of transport, to the final phase of destruction and recycling. During its useful life, different resources are used (e.g. lithium or cobalt in the manufacturing of batteries or natural gas to produce the energy stored in them), and environmental impacts occur (e.g. derived from the emissions when vehicle transits).

There are two traditional approaches to understand and quantify the environmental impact of any kind of vehicle technology: Life Cycle Assessment (LCA) and Well-to-wheel model (WTW). LCA is a tool to assess in a quantified way the environmental impact of manufacturing and using products [2]. In this way, the amount of material used, the energy consumed, the emissions and waste from all the entire life of a vehicle can be analysed in detail [3]. In addition, this tool has the potential to identify the critical processes from an environmental point of view, e.g. in the case of EVs the battery both manufacturing and end of life [4].

Well-to-Wheel (WTW) model zooms on a specific part of the LCA approach, which are the 'energy consumption' and 'Greenhouse Gas emissions' during the life of a vehicle, as shown in Fig. 1. Therefore, this second analyse does not include the energy or emissions during the manufacturing and end of life of the vehicles, nor the impact of vehicles in health or resources use. This analyse is structured into two stages: well-to-tank (WTT) and tank-to-wheel (TTW). The first stage, WTT, quantifies the energy required and the GHG emissions resulting from the production, transport and distribution of road transportation fuels, from the source to the fuel pump or plug. The second stage is TTW, and focuses on the vehicle, quantifying the emissions and energy use during driving, and therefore considering the efficiency of different powertrains. The resulting emissions and energy used for any technology can be obtained as the addition of both values $WTW = WTT + TTW$.

Both LCA and WTW analyses require first a detailed inventory of the use of material/resources and understanding of processes, and also a detailed calculation of indicators. LCA analysis focus on the use of material resources such as energy,

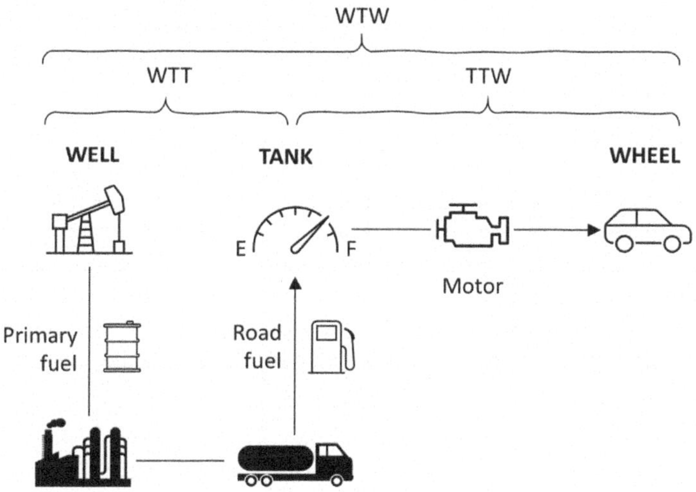

Fig. 1 Well-to-wheel analyses

emissions and waste, and the calculation of emissions and extraction of resources, to obtain the so-called environmental characterization factors [5]. This calculation in LCA process is also known as ReCiPe, which is based on the calculation of 18 midpoint indicators and 3 endpoint indicators. Figure 2 shows aggregated categories where it is necessary to analyse the impact, where the intermediate indicators analyse specific environmental problems, such as water use or acidification of the land, while the final indicators show the impact in three areas: effect on human health, biodiversity and lack of resources.

From a legal standpoint, the idea of life cycle should also be taken into consideration when governing environmental impacts and protection. For instance, at the international level it should be considered when regulating the international trade of pieces or other components (fuel included) related to the manufacture, use and disposal of vehicles.

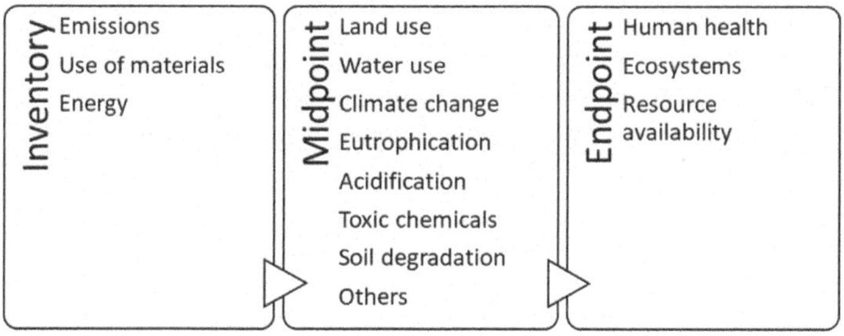

Fig. 2 ReCiPe [6] impact categories and endpoints

Focusing at the European level, Directive 2008/98/EC of the European Parliament and of the Council of 19 November 2008, on waste, includes the concept of life cycle in several articles:

- As a justification for Member States to depart from the waste hierarchy to deliver the best overall environmental outcome in relation to specific waste streams, taking into consideration the general environmental protection principles of precaution and sustainability, technical feasibility and economic viability, protection of resources as well as the overall environmental, human health, economic and social impacts (article 4.2).
- As part of the obligations that may be imposed on producers of certain products, when applying the 'extended producer responsibility scheme' (articles 8.2 and 8a.4.b).

The Communication from the Commission to the European Parliament, the Council, the European Economic and Social Committee and the Committee of the Regions 'Closing the loop—An EU action plan for the Circular Economy' (Brussels, 2 December 2015, COM [7] 614 final), must also be mentioned. This Communication refers to the transition to a more circular economy, where the value of products, materials and resources is maintained in the economy for as long as possible, and the generation of waste minimized.

Before this Communication, the Commission Recommendation 2013/179/EU of 9 April 2013, on the use of common methods to measure and communicate the life cycle environmental performance of products and organizations, had been published in the Official Journal of the European Union of 4 May 2013, L 124. This Recommendation promotes the use of the environmental footprint methods in relevant policies and schemes related to the measurement or communication of the life cycle environmental performance of products or organizations.

The rest of this chapter details the environmental impact of EVs based on the LCA and WTW models, covering the three main stages (manufacturing, use and end of life). A discussion on how regulation in Europe is covering these topics will be presented. It must be advanced that the environmental legislation in Europe is complete and covers all aspects that may refer to these three stages. However, it must adapt, as deemed appropriate at all times by the legislator, to the specific cases of the ICEs and the EVs according to the technologies and the requirements of environmental protection are evolving. The key issue is to apply the life cycle approach that has been pointed out, so a comprehensive and coherent environmental protection is ensured, beyond partial or poorly informed opinions.

2 Environmental Effects in the Manufacturing Process of EVs

The quantification of the environmental impact of manufacturing vehicles requires a comprehensive list of elements and processes, including three levels of disaggregation: (1) components, which are integrated into (2) subsystems and in turn are added into (3) units.

The main elements associated with the manufacture of the EV are summarized in Fig. 3, where the production of the different components and their assembly require the use of raw materials (steel, aliminium, copper, etc.), energy for manufacturing (electricity and heat), other materials (water, chemicals, etc.) and transportation of elements among assembly factories [5]. When analysing vehicles, regardless of technology, two large units can be differentiated: the structure of the car and the power train. Nowadays, it can be assumed that the structure and interior of the car—iron and plastics—is independent of the propulsion technology used, therefore there will be no difference between the conventional ICE and EVs regarding the main body [8]. Anyway, this is expected to change in near future, as the distribution of the weight in EVs is different: an important weight in conventional ICE vehicles is the power train, located in the front of the vehicle, while the main weight in EVs is in the battery pack, which is distributed on the floor of the vehicle. In addition, EVs will become autonomous in near future, and these vehicles will even change how passengers are seated, usually face to face [9]. To guarantee the safety of people in the car, under the event of a car crash, both the new distribution of weights and passenger positions will require new designs in car structures, and therefore differentiate from the structure of traditional ICE vehicles.

The power train in an EV consists of the battery pack and the electric motor. The battery of an EV is the critical unit from an environmental point of view, in comparison with traditional ICE vehicles, and therefore a detailed analysis is required. The battery back can be structured in 4 subsystems: cooling system, cells, structure and the

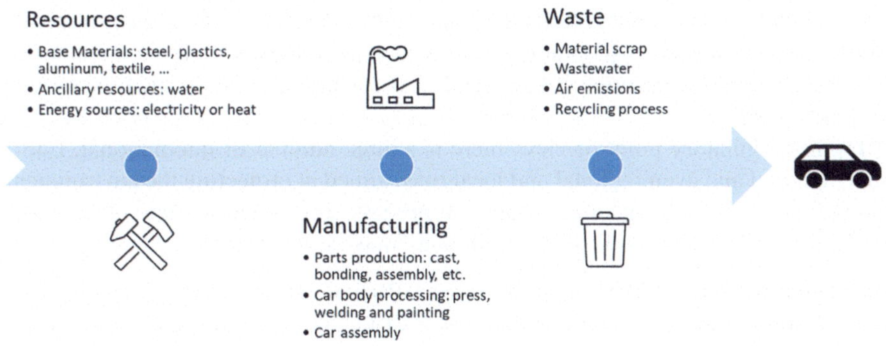

Fig. 3 Main elements in the production phase of a VE

Battery Management System. A typical distribution of weights among these subsystems is 4.1%, 60.1%, 32.1% and 3.7%, respectively [10], which also corresponds to their environmental impact importance.

The smallest elements in a battery are the cells, which are packed in modules, connected in series and parallel to build the battery pack (e.g. the Nissan Leaf has 192 cells packed in 48 modules). The basic components are the anode and cathode; the anode is the element where the oxidation reaction (loss of electrons) occurs, and the cathode is the element where the reduction reaction occurs (acquisition of electrons). Lithium-Ion batteries are currently a standard for most EVs. The cathode composition of these batteries can be very diverse: with Manganese $LiMn_2O_4$ (LMO), Iron Phosphate $LiFePO_4$ (LFP), Cobalt Li (NiCoAl) O_2 or mixture of several Li (NixCoyMnz) or O_2 (NCM), among others [11]. On the other hand, anodes are usually built with carbon or lithium titanate, to guarantee stability and safety. Lithium and Cobalt are the most relevant materials, from the environmental availability points of view, which directly impact its price. In 2018, the largest producer of Lithium was Australia (27.2 tonnes), followed by Chile (16 tonnes) and China (8 tonnes). On the other hand, Congo was the leading cobalt producing country (90 million tonnes) followed by Russia and Cuba (5 million tonnes), although Cobalt refineries are located in other regions such as China and Europe.

The second element of the power train is the electric motor. There are two technologies of electric motors used for the traction of EVs: Permanent Magnet and induction motors, used in 83% and 11% of EVs, respectively, leaving 6% of sales for EVs with both technologies [12]. Permanent Magnet motors are typically used because of their higher power density and efficiency compared to induction motors (for an average 50 kW permanent magnet motor is 30% lighter than a conventional induction motor). These improved characteristics can cost up to 4 times the conventional induction motor [13]. From the constructive point of view both motors consist of a cast iron or aluminium structure and copper or aluminium windings; the main difference between the two technologies is that synchronous motors additionally need permanent magnets. These magnets are constructed using so-called ‘rare earths’, such as neodymium. The world’s leading copper producer was Chile (5.5 Mton in 2017), followed by Peru and China . Regarding rare earths, the extraction of these minerals is concentrated in certain geographical areas such as China and in limited quantities [14], which makes there price volatile, and whose availability in the medium term is not guaranteed.

From a regulatory point of view, there is a huge number of international, European, national and even regional and local rules aimed at protecting the environment that will apply to each specific production process. Just to show some of the most important at the European level, the following may be mentioned:

(i) Directive 2010/75/EU of the European Parliament and of the Council of 24 November 2010, on industrial emissions (integrated pollution prevention and control). This Directive lays down rules on integrated prevention and control of pollution arising from industrial activities, as well as rules designed to prevent or, where that is not practicable, to reduce emissions into air, water and land and to prevent the generation of waste, in order to achieve a high level of protection of the environment taken as a whole (article 1). This Directive applies to the activities set out in Annex I (article 10) and to other specific activities referred to in Chaps. III to VI of the Directive (article 2.1), but not to research activities, development activities or the testing of new products and processes (article 2.2).

Among the activities subject to the Directive the following within Annex 1 may be cited: combustion of fuels in installations with a total rated thermal input of 50 MW or more (Section 1.1), certain installations for the processing of ferrous metals (Section 2.3), installations for the manufacture of glass including glass fibre with a melting capacity exceeding 20 tonnes per day (Section 3.3), installation for the production of organic chemicals and of inorganic chemicals (Sections 4.1 y 4.2), installations for the disposal or recovery of hazardous waste with a capacity exceeding 10 tonnes per day involving one or more of the following activities (Section 5.1) and installations for the surface treatment of substances, objects or products using organic solvents, in particular for dressing, printing, coating, degreasing, waterproofing, sizing, painting, cleaning or impregnating, with an organic solvent consumption capacity of more than 150 kg per hour or more than 200 tonnes per year (Section 6.7).

(ii) Directive 2011/92/EU of the European Parliament and of the Council of 13 December 2011, on the assessment of the effects of certain public and private projects on the environment. This Directive applies to the assessment of the environmental effects of public and private projects which are likely to have significant effects on the environment (article 1.1).

Among others, certain projects referred to the energy industry (Section 2 of Annex I), integrated works for the initial smelting of cast iron and steel (Section 4.a) and the mining industry (Section 19), are subject to this Directive.

(iii) Directive 2008/50/EC of the European Parliament and of the Council of 21 May 2008, on ambient air quality and cleaner air for Europe.

This Directive lays down measures aimed at, inter alia, defining and establishing objectives for ambient air quality designed to avoid, prevent or reduce harmful effects on human health and the environment as a whole; assessing the ambient air quality in Member States on the basis of common methods and criteria; and obtaining information on ambient air quality to help combat air pollution and nuisance (article 1).

The Directive refers to pollutants such as sulphur dioxide, nitrogen dioxide and oxides of nitrogen, particulate matter, carbon monoxide and ozone.

(iv) Directive 2000/60/EC of the European Parliament and of the Council of 23 October 2000, establishing a framework for Community action in the field of water policy. The purpose of this Directive is to establish a framework for the protection of inland surface waters, transitional waters, coastal waters and groundwater which, among others, prevents further deterioration and protects and enhances the status of aquatic ecosystems; promotes sustainable water use based on a long-term protection of available water resources and aims at enhanced protection and improvement of the aquatic environment.

Among others, the Directive contains rules on environmental objectives, protected areas, recovery of costs for water services and strategies against pollution of water.

(V) Directive 2003/87/EC of the European Parliament and of the Council of 13 October 2003, establishing a system for greenhouse gas emission allowance trading within the Union. This Directive establishes, among others, a system for greenhouse gas emission allowance trading within the Union in order to promote reductions of greenhouse gas emissions in a cost-effective and economically efficient manner (article 1).

This Directive applies to emissions from the activities listed in Annex I (for instance, combustion of fuels in installations with a total rated thermal input exceeding 20 MW—except in installations for the incineration of hazardous or municipal waste; metal ore—including sulphide ore—roasting or sintering, including pelletization; and production of primary aluminium) and greenhouse gases listed in Annex II (carbon dioxide, methane, nitrous oxide, hydrofluorocarbons, perfluorocarbons and sulphur hexafluoride).

In addition to these rules, and for the purposes of this article, it is worth mentioning Directive 2000/53/EC of the European Parliament and of the Council of 18 September 2000, on end of life vehicles, that lays down measures which aim, as a first priority, at the prevention of waste from vehicles and, in addition, at the reuse, recycling and other forms of recovery of end of life vehicles and their components so as to reduce the disposal of waste, as well as at the improvement in the environmental performance of all of the economic operators involved in the life cycle of vehicles (article 1).

Among other things, and in order to promote the prevention of waste, this Directive imposes on Member States the obligation to encourage (article 4.1):

- vehicle manufacturers, in liaison with material and equipment manufacturers, to limit the use of hazardous substances in vehicles and to reduce them as far as possible from the conception of the vehicle onwards, so as in particular to prevent their release into the environment, make recycling easier, and avoid the need to dispose of hazardous waste;
- the design and production of new vehicles which take into full account and facilitate the dismantling, reuse and recovery, in particular the recycling, of end of life vehicles, their components and materials.

Member States must also ensure that materials and components of vehicles put on the market after 1 July 2003 do not contain lead, mercury, cadmium or hexavalent chromium (other than in cases listed in Annex II of the Directive, under the conditions specified therein—article 4.2).

3 Environmental Effects in the Use of EVs

The environmental impact of EVs during their useful life can be classified in direct emissions due to wheel-to-road friction (also called non-exhaust emissions such as small particles) and maintenance, and an indirect impact on the use and transport of resources to provide electric charging of the battery, as indicated in Fig. 4.

For direct emmisions in the process of combustion, ICE vehicles produce a set of gases that have an impact on the air quality, while EVs do not produce these gases. Additionally, the movement of the vehicle in a road increases the generated PM_{10} and $PM_{2.5}$ due to factors such as brake wear, road wear, tyre wear and road dust resuspension. These emissions are called non-exhaust emissions and are almost proportional to the mass of the vehicle. Currently, due to the weight of the battery pack EVs are around 20% heavier than the corresponding diesel or gasoline vehicles, resulting in the fact that EVs emit similar particulates compared to modern ICE vehicles; assuming that EVs have no particle emissions due to exhaust or brake wear, being heavier will result in slightly higher values for road wear and resuspension of particles. Some authors quantified a reduction of 1–3% in $PM_{2.5}$ for EVs and no reduction in PM_{10} [15].

A second direct environmental impact occurs during the maintenance of the vehicle that can be predictive and corrective, although it has a much lower impact than the previous circulation effect. As EVs have almost 50% less mechanical components compared to traditional vehicles, this will directly imply a higher reliability and lower maintenance. The predictive maintenance process for any vehicle will require a periodic review of elements such as brake and cooling fluids, as well as the cabin air filter; however, for EVs it is not necessary to include oil changes (engine or gearbox), as well as fuel filters, belts and other ancillary elements. According to some studies, the maintenance of the EV is almost 40% cheaper than their diesel or gasoline counterpart [16]. On the other hand, corrective maintenance, associated with possible breakdowns, is expected to be significantly lower in EVs, since they are simpler and more robust than ICE vehicles, avoiding traditional problems associated

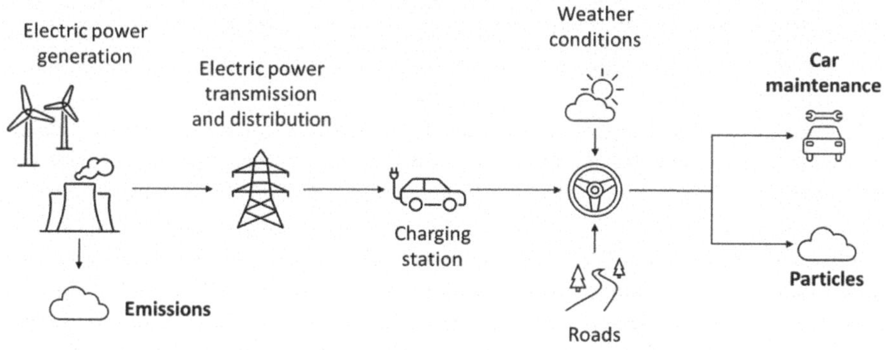

Fig. 4 Emissions associated with the use of the VE

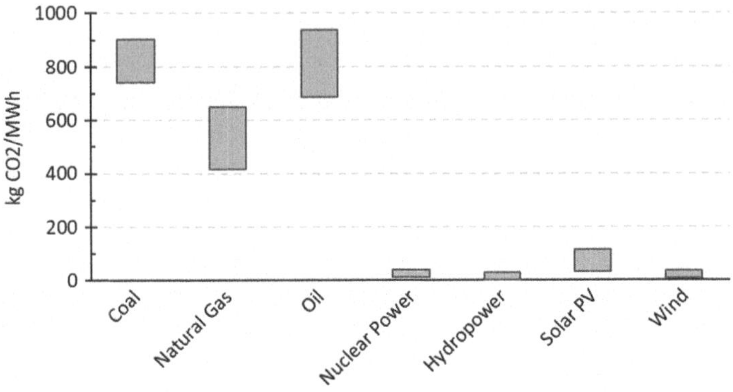

Fig. 5 Life cycle emission factors for electricity generation from selected technologies. *Source* Own elaboration

with the timing belt breaks or problems in the injectors. However, embedded systems and software designs are becoming the main elements for EV reliability [17].

Although EVs do not have direct exhaust emissions, they could have relevant indirect emissions associated with the generation of electricity required to charge batteries. The process of electric generation requires energy and have emissions. This process includes the extraction and transportation of raw material—such as gas or coal—to the generation power plants, the conversion to electricity, and the transmission and distribution of the electricity through the networks down to the charging poles. From the energy point of view, the production of electrical energy has a typical efficiency in the range of 30%, while the transport and distribution of electrical energy represents losses of around 15%. Electricity generation emission factors include the greenhouse gas carbon dioxide (CO_2) and additional gases which are not greenhouse but are considered as critical pollutants sulphur dioxide (SO_2), and nitrogen oxides (NO_x). Figure 5 compares the emissions of different electric generation technologies, using a Life Cycle Analysis, that includes not only the use of resources but also the manufacturing of the power plant components. From this figure it is clear that renewable generation can provide up to 100 times less exhausting gases (both GHG and other pollutants) compared with traditional fossil fuels [18]. Then, the emissions associated with the transition to electro-mobility in each country will mostly depend on its power generation mix, as shown in Fig. 6 [13].

WTW analysis also helps to characterize for different traction technologies global energy use and compare local emissions (TTW). Table 1 compares the efficiency and CO_2 emissions for different technologies of a medium size utility car with an average consumption of 15 kWh/100 km, supplied by a partially green energy mix [21, 22]. Results in TTW show that BEV together with Fuel Cell Vehicles are clearly the ones that provide the lowest impact in air quality locally compared with fossil fuels. From a country or regional perspectives, it is quite relevant the energy required for the transportation sector (WTT), as it affects regional imports; in this section, EVs are still the best option with close to overall 50% efficiency, followed by Hybrid EVs.

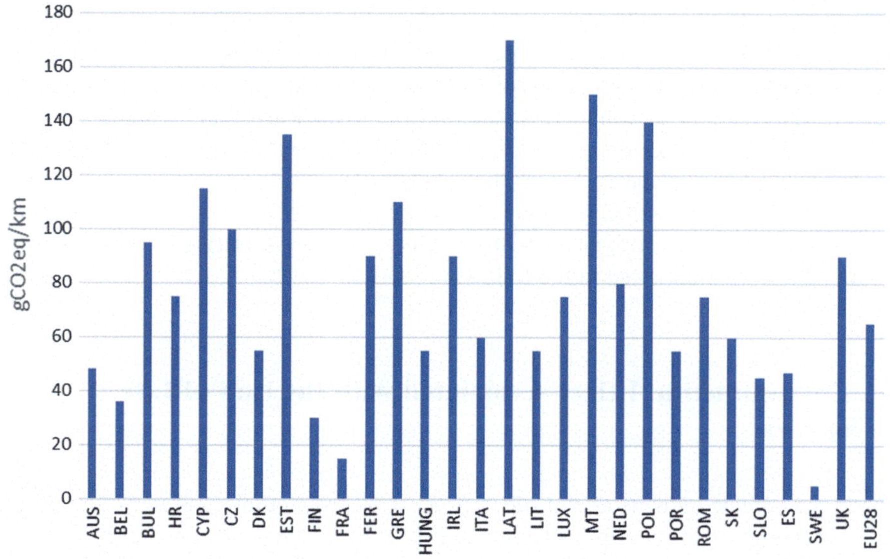

Fig. 6 WTW CO_2 emissions for electric vehicles in different EU Member States for 2013 generation mix [19, 20]

Table 1 WTW comparison of emissions and efficiency of different traction technologies

	WTT		TTW		WTW	
	Efficiency (%)	gCO_2/km	Efficiency (%)	gCO_2/km	Efficiency (%)	gCO_2/km
Gasoline EURO 6C	81	54.04	18	192	15	246
Diesel EURO 6C	86	42.96	21	170	18	213
GLP	88	47.77	15	181	13	228
GNC	89	25.82	18	162	16	188
FCEV	58	95	42	0	25	95
HEV	81	54.04	44	109	35	163
BEV	61	51.62	77	0	47	52

In relation to the rules on emissions from vehicles, Regulation (EU) 2019/631 of the European Parliament and of the Council of 17 April 2019, setting CO_2 emission performance standards for new passenger cars and for new light commercial vehicles, must be cited. This Regulation establishes CO_2 emissions performance requirements for new passenger cars and for new light commercial vehicles in order to contribute to achieving the Union's target of reducing its greenhouse gas emissions, as laid down in Regulation (EU) 2018/842, and the objectives of the Paris Agreement and to ensure the proper functioning of the internal market (article 1). It also establishes that the Commission shall no later than 2023 evaluate the possibility of developing a common Union methodology for the assessment and the consistent data reporting of the full life-cycle CO_2 emissions of passenger cars and light commercial vehicles that are placed on the Union market, and transmit to the European Parliament and to the Council that evaluation, including, where appropriate, proposals for follow-up measures, such as legislative proposals (article 7.10).

Other rules mentioned above may also be taken into consideration when dealing with the use of EV, such as those on the production of waste derived from such use. Some others are mentioned in the following section.

4 Environmental Effect Associated with the End of Life of EVs

When the vehicle ends its useful life, a process of disassembly and recycling of its different components begins, as shown in Fig. 7. In the process of recycling of EVs the valuable components are those that have metal parts that can be found in the body of the vehicle (mainly steel), battery pack (such as nickel, cobalt, manganese) and power train (aluminium and copper) [5]. The processes for recycling each metal are usually very energy intensive.

Regarding the battery, when its capacity reduces below 80% of the original value it is not appropriate for traction applications, and therefore needs to be retired. However, the battery can still be used for other applications such as stationary electrical storage devices. This application will be integrated into the power system, typically in low and medium voltages, to store energy integrated into a photovoltaic or wind power plants [23], to provide different services, such as reducing or eliminating congestions in the network, support voltages. Lifespan estimation for this second life of the battery can go from 6 to 30 years depending on the services they procured [24].

Finally, in case the capacity of the battery is exhausted with less than 20% of the original capacity value, the process starts from a pyrometallurgical furnace in

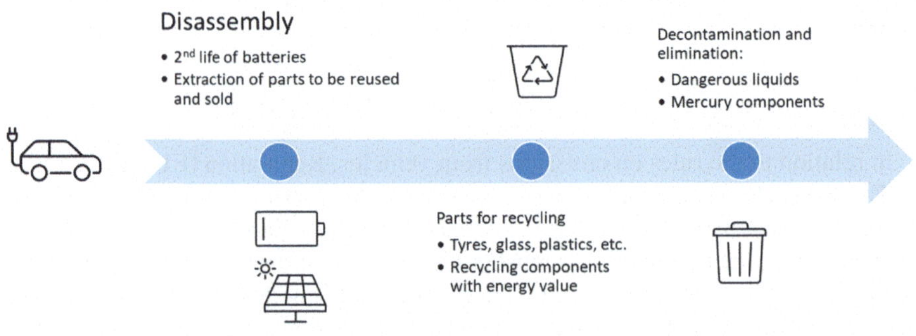

Fig. 7 Main processes at the end of the life cycle of the VE

order to recover cobalt and nickel. The slag is mainly composed of compounds of Aluminium, Silicon, Calcium, Iron and also Lithium Oxide. This slag can be used as an additive in construction or cement [25].

According to the Directive 2000/53/EC mentioned above, and among other rules, Member States shall take the necessary measures to ensure:

- that economic operators set up systems for the collection of all end of life vehicles and, as far as technically feasible, of waste used parts removed when passenger cars are repaired (article 5.1);
- the adequate availability of collection facilities within their territory (article 5.1);
- that all end of life vehicles are transferred to authorized treatment facilities (article 5.2);
- that all end of life vehicles are stored (even temporarily) and treated in accordance with the waste hierarchy and the general requirements laid down in article 4 of Directive 2008/98 cited above, and in compliance with certain minimum technical requirements set out in Annex I to the Directive 2000/53/EC (article 6.1);
- that any establishment or undertaking carrying out treatment operations fulfils at least the certain obligations as set out in the Directive, such as the following (article 6.3): (a) end of life vehicles shall be stripped before further treatment in order to reduce any adverse impact on the environment; (b) hazardous materials and components shall be removed and segregated in a selective way so as not to contaminate subsequent shredder waste from end of life vehicles; and (c) stripping operations and storage shall be so as to ensure the suitability of vehicle components for reuse and recovery, and in particular for recycling.
- that producers provide dismantling information for each type of new vehicle put on the market within six months after the vehicle is put on the market (article 8.3).
- (without prejudice to commercial and industrial confidentiality) that manufacturers of components used in vehicles make available to authorized treatment facilities, as far as it is requested by these facilities, appropriate information concerning dismantling, storage and testing of components which can be reused (article 8.4).

Member States shall set up a system according to which the presentation of a certificate of destruction is a condition for deregistration of the end of life vehicle (article 5.3). The Directive also sets certain reuse and recovery targets (article 7).

5 Overall Environmental-Global Impact of EVs

Once the impacts in the whole life cycle of EVs have been detailed in the previous sections, Fig. 8 summarizes these standarized impacts, with a focus in 3 relevant midpoints from the 18 defined in the ReCiPe Model: Global warming and depletion of mineral and fossil resources) [26]. Results compare different traction technologies in the utility segment (ICE and EVs) and two Lithium battery technologies for EVs (Li-NCM and Li-$FePO_4$). According to these results, the environmental impact of

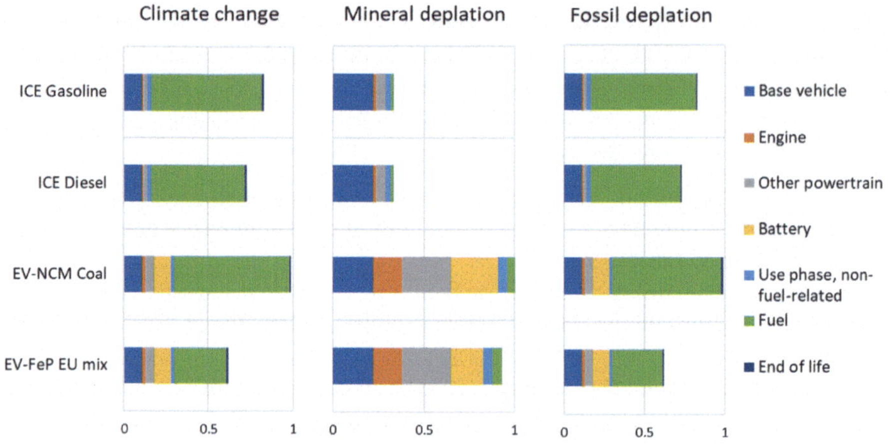

Fig. 8 Standardized vehicle impact (based on [26])

EVs is concentrated in the manufacturing phase of the batteries and the motor, while the impact of ICE-based vehicles is more relevant in their use phase. In case of EVs supplied with an average European generation mix, we would able to reduce the impact on global warming by up to 24% compared to ICE vehicles, and also Li-$FePO_4$ batteries have lower impact compared to Li-NCM [27].

The transition from conventional vehicles into electric mobility will have an impact on the use of natural resources and raw materials, and hence will certainly have international geopolitical effects. In the manufacturing phase, EVs (including hybrid and battery) need new raw materials, such as cobalt, lithium, rare earths and copper are required in high amounts. This will allow new players to enter into supply chain, such as Chile, Congo and China.

But it is even more important the use phase, where EVs will typically run with autochthonous energy sources such as renewables or coal in Germany or the UK. This represents a change with respect to current status based on the oil supply chain (diesel or gasoline and lubricants), where traditional oil production is concentrated in North America, Russia and the Middle East.

Therefore, as it is shown in Fig. 9, it is expected a change in the role of countries and regions as mayor suppliers of raw material for road transportation. This transition requires to adequality measure the possible negative environmental effects in the different areas to regulate in such a way that these effects could be eliminated or at least compensated. Although the estimations of EV sales in the coming years are relevant, the change of the vehicle fleet will be progressive, as it is expected that the upgrading rate of cars per year remains the same, typically 4%, leaving up to 25 years for a complete vehicle technological change.

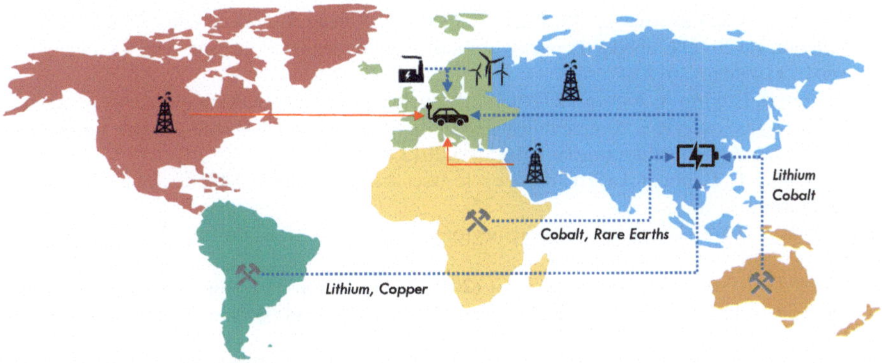

Fig. 9 Geopolitical effect of technology change in transportation (red: traditional supply chain, blue: EVs supply chain)

References

1. Colvile RN, Hutchinson EJ, Mindell JS, Warren RF (2001) The transport sector as a source of air pollution. Atmos Environ 35(9):1537–1565. ISSN 1352-2310, https://doi.org/10.1016/S1352-2310(00)00551-3
2. Baumann H, Tillman A-M (2004) The Hitch Hiker's Guide to LCA. In: An orientation in life cycle assessment methodology and application. Lund, Sweden, Studentlitteratur
3. Curran MA (1996) Environmental life-cycle asessment. McGraw Hill Professional, New York
4. Clift R (2006) Sustainable development and its implications for chemical engineering. Chem Eng Sci 61:4179–4187. https://doi.org/10.1016/j.ces.2005.10.017
5. Del Duce A et al (2013) ELCAR guidelines for the LCA of electric vehicles. Deliverable 2.1 of E-mobility life cycle assessment recommendations, FP7 EU Project
6. ReCiPe 2016 v1.1 A harmonized life cycle impact assessment method at midpoint and endpoint level Report I: characterization RIVM Report 2016-0104a
7. https://eur-lex.europa.eu/resource.html?uri=cellar:8a8ef5e8-99a0-11e5-b3b7-01aa75ed71a1.0012.02/DOC_1&format=PDF
8. Ecoinvent Centre (2010) Ecoinvent data and reports, v2.2, Dübendorf Switzerland, Swiss Centre for Life Cycle Inventories
9. Hirz M, Rossbacher P (2018) Integrated approach supporting virtual conception of passenger ergonomics in automated driving cars. In: Conference: 9th international conference on applied human factors and ergonomics—AHFE 2018 At, Orlando, USA
10. Ellingsen LA-W, Majeau-Bettez G, Singh B et al (2014) Life cycle assessment of a lithium-ion battery vehicle pack. J Ind Ecol 18:113–124. https://doi.org/10.1111/jiec.12072
11. Goodenough JB, Park KS (2013) The Li-ion rechargeable battery: a perspective. J Am Chem Soc 135(4):1167–1176. https://doi.org/10.1021/ja3091438
12. IDTechEx (2014) Electric motors for electric vehicles 2013–2023: forecasts, technologies, players
13. Burnell (2013) Performance/cost comparison of induction-motor & permanent-magnet-motor in a hybrid electric car Malcolm Burwell. In: International Copper Association James Goss, Mircea Popescu—Motor Design Ltd, Tokyo
14. Le Petit Y (2017) Electric vehicle life cycle analysis and raw material availability. Trans Environ
15. Timmers VRJH, Achten PAJ (2016) Non-exhaust PM emissions from electric vehicles. Atmos Environ 134. https://doi.org/10.1016/j.atmosenv.2016.03.017
16. Raustad R (2017) Electric vehicle life cycle cost analysis. EVTC Electric Vehicle Transportation Center

17. Chakraborty S, Lukasiewycz M, Buckl C, Fahmy S, Chang N, Park S, Kim Y, Leteinturier P, Adlkofer H (2012) Embedded systems and software challenges in electric vehicles. https://doi.org/10.1109/date.2012.6176508
18. Turconi R, Boldrin A, Astrup TF (2013) Life cycle assessment (LCA) of electricity generation technologies: overview, comparability and limitations Published. Renew Sustain Energy Rev Link to article. https://doi.org/10.1016/j.rser.2013.08.013
19. Ager-Wick L, Hammer A (2017) Life cycle assessment of electric vehicles. In: 12th Concawe symposium
20. JRC (2017) CoM default emission factors for the member states of the European Union dataset version 2017
21. Notter DA, Gauch M, Widmer R et al (2010) Contribution of Li-ion batteries to the environmental impact of electric vehicles. Environ Sci Technol 44:6550–6556. https://doi.org/10.1021/es903729a
22. Tagliaferri C et al (2016) Life cycle assessment of future electric and hybrid 1 vehicles: a cradle-to-grave systems engineering approach. Chem Eng Res Des 112:298–309 (Elsevier)
23. Heymans C, Walker SB, Young SB, Fowler M (2014) Economic analysis of second use electric vehicle batteries for residential energy storage and load-levelling. Energy Policy 71:22–30. ISSN 0301-4215
24. Casals LC, García BA, Canal C (2019) Second life batteries lifespan: rest of useful life and environmental analysis. J Environ Manag 232:354–363. ISSN 0301-4797
25. Vadenbo C (2009) Prospective environmental assessment of lithium recovery in battery recycling
26. Hawkins TR et al (2012) Comparative environmental life cycle assessment of conventional and electric vehicles. J Ind Ecol 17(1). https://www.pre-sustainability.com/download/Report_ReCiPe_2017.pdf
27. Messagie M (2014) Life cycle analysis of the climate impact of electric vehicles. Transp Environ
28. Moro A, Lonza L (2018) Electricity carbon intensity in European member states: impacts on GHG emissions of electric vehicles. Transp Rese Part D: Transp Environ 64:5–14

A Macroeconomic Contribution: Extended Environmental Input–Output Analysis

Manuel Ordóñez Ríos, Angel Arcos-Vargas, José Manuel Cansino Muñoz-Repiso, and Rocío Román Collado

Abstract An environmentally extended Input–Output (IO) model is used to estimate the impact of the increased use of electric cars (EV) on production, Gross Value Added, employment, and greenhouse gas emissions. The year taken as a reference is 2030 and the analysis is carried out for a sample of 29 countries that include the EU 28 and Norway. The reference databases for the IO model correspond to 2014 and the four sectors mainly impacted by the introduction of EV are coking plants and oil refining, the manufacture of motor vehicles, commerce and repair of motor vehicles, and the supply of electric power The results vary significantly between countries although the greatest impacts appear in those located in Eastern Europe.

1 Introduction

In the transition toward a neutral economy in greenhouse gas emissions (GHG), electric cars (EV) are the main driving force behind the decarbonization of the mobility and transport sectors. GHG emissions associated with transport represent more than a quarter of the total GHG emissions of the European Union (EU). Their evolution also shows a different behavior from other traditionally polluting sectors. Thus, for example, while energy production and industry have reduced their emissions since 1990, transport emissions have increased [2]. They represent more than a quarter of the total GHG emissions in the EU. Particularly, road transport, such as cars, trucks, trucks, and buses, produces more than 70% of the total GHG emissions from transport while the rest comes mainly from sea and air transport. All of the above reinforces the role of EV in this sector's decarbonization process.

Following Cansino and Yñiguez [1], EV are vehicles that comply with the definition provided in Directive 2007/46/EC of the European Parliament and of the Council

M. O. Ríos · A. Arcos-Vargas (✉) · J. M. C. Muñoz-Repiso · R. R. Collado
University of Seville, Seville, Spain
e-mail: aarcos@us.es

J. M. C. Muñoz-Repiso · R. R. Collado
Universidad Autónoma and de Chile, Santiago, Chile

A. Arcos-Vargas (ed.), *The Role of the Electric Vehicle in the Energy Transition*, Green Energy and Technology, https://doi.org/10.1007/978-3-030-50633-9_7

of 5 September 2007. This Directive establishes a framework for the approval of motor vehicles and their trailers, as well as the systems, components, and separate technical units intended for such vehicles.

Over the years, the EU has taken various actions to support electric mobility. The main UE pillars supporting the use of EV are the Renewable Energy Directive 2009/28/EC, the Fuel Quality Directive 2009/30/EC, the Clean Vehicle Directive 2009/33/EC, the Regulations setting CO_2 standards for passenger cars (Regulation No 443/2009) and light commercial vehicles (Regulation No 510/2011), the Directive 2009/28/EC by the European Parliament and the Council dated 23/04/2009 for the promotion of energy from renewable sources, the Commission's Communication on a European alternative fuels strategy [3], and the Directive 2014/94/EU on the deployment of alternative fuels infrastructure in Europe.

The replacement of traditional combustion vehicles with EV will cause important changes in electricity requirements, in the volume of GHG emissions and in employment. All these changes will impact on an industry whose relevance has recently been highlighted by the European Parliament itself, pointing out that the European automotive industry is, at the same time, an important link in the industrial production chain and a fundamental factor of competitiveness, growth, and employment for Europe [3]. For this reason, the size that the car industry has in some EU countries justifies an analysis such as that developed in this chapter, in which the impact that a change in the transport sector aimed toward the electromobility is estimated.

This chapter calculates the impacts caused by the penetration of EV in three possible scenarios in the 28 EU countries, before the United Kingdom left, also including Norway. The penetration scenarios proposed are 10, 20, and 30%, with 2030 as the reference year. These scenarios have a central reference value as contemplated in the Paris Declaration on Electro-Mobility and Climate Change (20%). From a large and rich database, the calculations are derived from an Input–Output (IO) model of quantities that are subsequently extended to analyze both energy requirements and the impact on GHG emissions. The analysis is performed disaggregated for 55 economic sectors and enables the impact of the introduction of EV on Gross Value Added (GVA), emissions, and employment to be calculated.

2 Database

The main statistical information used to prepare this chapter has been the national TIOs published by the World Input–Output Database, which refer to the year 2014 [14, 16]. These tables show data related to 55 economic sectors and branches of activity of a set of 29 countries. The total employment data have been taken from this same database for each of the branches considered [15].

Most of the other data come originally from Eurostat; GHG emissions [4], the value of industry production [5], that of commerce [6], the savings rate [7], international trade [8], the vehicle fleet [9], electric consumption [10], and electricity production by type of fuel [11].

Together with the above, international trade data have been taken for Norway, specifically the exports of goods to the EU and to countries outside the EU, from Statistics Norway [13].

3 Methodology

The Input–Output Tables (TIO) are a statistical base where all the production activity of an economy with the resources used and the jobs necessary for the production of goods and services is exhaustively described. A TIO collects all the interactions that appear between the different economic sectors or branches of activity and that of these with the primary factors (labor and capital remuneration) and with the institutional sectors through the final demand. These are double-entry tables showing the set of economic transactions that have taken place between the economic agents during a certain period of time, which is normally the calendar year.

In addition to this statistical nature, the TIOs have a strong analytical character. They are used as a tool to analyze the results derived from the application of specific economic policies, whether for the whole economy of a country or, specifically, for each of the economic sectors. The main Input–Output models used in the economic analysis are defined from the TIOs.

The methodology used in this chapter is based on Leontief's model [12] that allows evaluating the economic impact which the activities of economic agents have on an economy's production sectors. This methodology is based on the information contained in the TIOs. The methodology is specified in the fundamental equation of the Input–Output model which indicates that the production of each sector depends on the final demand:

$$x = (I - A)^{-1} \cdot f \tag{1}$$

where x is a vector $n \times 1$, which signifies the value of the total production of each production sector, where n is the number of production sectors; I is the identity matrix with dimension $n \times n$; A is the matrix of technical coefficients, $n \times n$, where each of its lij elements indicates the needs that a given sector has of the inputs of another sector per unit of production. Finally, f is a vector with dimension $n \times 1$, which represents the final demand, that is, the demand that economic agents make of each production sector considered. The matrix is called Leontief's inverse matrix and each of its elements are the so-called simple multipliers of the model and represent the amount of output that sector i must produce to increase the final demand of sector j by one unit or, also, the input needs of sector i that are necessary to manufacture a good unit by sector j.

Taking into account the previous expression, any variation that occurs in the final demand will lead to a variation in production due to the interrelations between the sectors, and which are considered by the inverse matrix of Leontief. The results obtained are the sum of the direct and indirect effects caused by the impact on demand.

In addition to analyzing the economic impact of the introduction of EV on production, the model allows calculating the impact on employment, gross value added, or emissions generated by production activities, among other variables. Taking as an example the impact on emissions, the first step is to define the matrix of emission coefficients ($\hat{E}$). This is a diagonal matrix of dimension $n \times n$ where each of the elements of the main diagonal is the ratio between the level of emissions (e_j) and the total output (X_j) for each production sector.

$$E_j = \frac{e_i}{X_j} \tag{2}$$

Once the diagonal matrix of emission coefficients is defined, it can be incorporated into the fundamental equation of the Leontief model (3) as follows:

$$e = \hat{E} \cdot X = \hat{E} \cdot (I - A)^{-1} \cdot f \tag{3}$$

where e is the emission vector of dimension $n \times 1$ in which each of its elements indicates the emissions originated by each production sector before a change in demand.

In the same way that a matrix of emission coefficients has been defined, this can be done with employment and with the added value and introduced into the model in a similar way to that in (3) in order to find out the impact that a given demand shock has on these variables.

To analyze the impact that the introduction of EV will have on the economies of the EU countries and Norway, the final demand vector has been modified. So the new demand vector includes the same values as the vector supplied by the TIOs except in the values corresponding to the four branches of activity that are considered to be directly affected by the introduction of EV (*Coke and oil refining industry, Manufacture of motor vehicles, Trade and Repair of Motor Vehicles, Electricity Supply*). For these branches, the initial values of the demand vector have been modified according to the three scenarios considered: 10, 20, and 30% penetration of electric cars with the time horizon of the year 2030. These scenarios are based on that contemplated in the Paris Declaration on Electro-Mobility and Climate Change (20%), adding a more moderate and a more ambitious one.

The *Coke and oil refining industry* will be affected due to the decrease in demand for petroleum products. The decrease that has been considered has been similar to the penetration coefficients of EV, that is, 10, 20, and 30%. Since oil refining accounts for 99.5% of the total production of the sector in the EU, it has been considered that the reduction in final demand for this branch affects the whole. However, this reduction will only apply to the institutional sectors: household consumption, nonprofit institutions serving households (NPISH), public consumption, and exports directed toward EU countries. In the latter case, exports are weighted according to the export coefficients shown in Table 2 of the annex. On the other hand, it is assumed that this negative impact on petroleum products is not compensated by higher exports to non-EU countries.

The *Manufacture of motor vehicles* branch of activity will be adversely affected by the introduction of EV due to the lower demand for EV maintenance services. It is estimated that EV maintenance is 80% lower than that of a traditional vehicle. With an EV penetration coefficient of 10%, this will mean that the maintenance of the vehicle fleet in parts and accessories will be reduced by 8%, since 90% of the vehicles will continue to have traditional maintenance (16 and 24% for the scenarios 2 and 3). On the other hand, as the manufacture of motor vehicles, trailers, and semitrailers activity includes three activity subbranches (motor vehicles; bodies, trailers, and semitrailers; and parts and accessories), the reduction will only be applied to the subbranch parts and accessories for each of the countries. This subbranch represented 28.9% of this productivity sector in 2014 in the EU as a whole, although its relative weight varies between countries as can be seen in Table 3 of the annex. The reduction will be applied to household consumption, nonprofit institutions serving households (NPISH), public consumption, variation of stocks, and exports directed toward EU countries.

The branch of activity *Trade and Repair of Motor Vehicles* is expected to suffer a negative impact as a result of the penetration of EV for the aforementioned reasons. In this case, a scenario similar to the previous case has been considered, assuming that EV maintenance is 80% lower than that corresponding to a traditional vehicle. In this case, this branch of activity encompasses four subbranches and the decrease has only been considered in two of them: maintenance and repair of motor vehicles and trade in motor vehicle accessories. In the same way, as for the previous branch, the reduction coefficients have been calculated and are those that are reflected in Table 4 of the annex. This considered decrease has been applied to household consumption, nonprofit institutions serving households (NPISH), public consumption, inventory variation, and exports directed toward EU countries.

The *Electricity Supply* will be the last branch affected by the modification of the final demand vector. Unlike the previous ones, in this case the result expected will not imply a decrease in consumption but an increase given the penetration of EV in society. For the three scenarios considered, the demand for the Production and distribution of electricity subbranch and the Electricity, gas, and water branch are those detailed in Table 5 of the annex. The increase has been applied to the following components of demand: household consumption, nonprofit institutions serving households (NPISH), public consumption, and exports directed toward EU countries.

On the other hand, the change in the budget constraint of the economic agents will produce a readjustment in their shopping baskets. The analysis assumes that the decrease in the economic agents' expenditure as a result of the introduction of EV (a lower expense in the maintenance and repair of vehicles and lower expenditure on petroleum products that compensate for the greater expenditure on electricity consumption) will mean an increase in the purchasing power of households and the NPISH. In our analysis, part of this increase in the purchasing power will be used for savings (see Table 6 in the annex) and the rest will be distributed proportionally to the weight that each component of the final demand of households had in the year 2014 (WIOD). This behavior can cause a rebound effect on consumption which will be referred to later.

Finally, the analysis assumes that the penetration of EV will mean a decrease in GHG emissions. To this end, the emission coefficient of the land transport sector has been reduced as it is the sector where it is estimated that it will have the greatest impact. The decrease in the emission coefficient applied has been calculated in a linear manner for the three scenarios considered: 10, 20, and 30%.

4 Results

Table 1 shows the results on production, GVA, employment, and the level of emissions caused by the introduction of EV in the three scenarios considered for the 29 countries analyzed.

The effect of the introduction of EV on production is reduced, ranging, for scenario 2, from an increase of 0.51% in Cyprus to a fall of 0.96% in Hungary. In the case of Cyprus, the rise can be explained by its sectoral structure, since the production value of the oil refining and manufacture of motor vehicles branches is minimal as can be seen in Fig. 1, so that this is barely affected. Something similar, although with a smaller increase in production, is observed in Latvia and Croatia. However, in these two countries the weight of the electricity sector is greater than in Cyprus.

However, the introduction of EV causes an increase in production in most sectors, being more important in the electricity sector, real estate activities, housing and catering and food, beverages, and tobacco. On the contrary, in the case of Hungary there is a drop-in production in most sectors, this being especially important in the manufacture of motor vehicles and petroleum refining. In Hungary, the fall observed in the manufacture of motor vehicles is due to the fact that the manufacture of parts and accessories accounts for about 43% of the total production of this branch of activity, representing 10% of the total production of the country, as can be seen in Fig. 1. Something similar appears in Czechia and Slovakia with an important weight in the vehicle manufacturing sector and, within this, the manufacture of parts and accessories. Also noteworthy is the decline in production observed in Belgium and Lithuania as a result of the significant weight of the oil refining sector, which in the case of Lithuania reaches 9.1% of the value of production.

The second of the variables studied is the gross value added at basic prices. In this case, the behavior of this economic magnitude before the introduction of EV in scenario 2, varies from one country to another. The range of variations is somewhat smaller than in the case of production, with extreme values for Hungary, with a decrease of 0.46%, and Cyprus, with an increase of 0.52%. The reasons must be sought, as in the case of production, in the weight of the economic sectors in the economy as a whole. Hungary has a high weight in the vehicle manufacturing sector in the GVA as a whole and, within this branch, the subpart manufacturing of parts and accessories attains about half of the GVA. In the case of Cyprus, the weight of the oil refining, manufacture of motor vehicles, and commerce and vehicle repair sectors is minimal, representing 1.9% of the total GVA.

Table 1 Effects on production, GVA, employment, and GHG emissions of EV penetration in the EU + Norway (%)

	Sce 1				Sce 2				Sce 3			
	Prod	GVA	Empl	Emis	Prod	GVA	Empl	Emis	Prod	GVA	Empl	Emis
Austria	−0.05	0.00	0.00	−0.47	−0.10	0.00	0.00	−0.93	−0.15	0.00	0.00	−1.40
Belgium	−0.44	−0.15	−0.15	−0.84	−0.87	−0.31	−0.30	−1.69	−1.42	−0.59	−0.57	−2.54
Bulgaria	0.06	0.18	0.14	−0.32	0.12	0.35	0.27	−0.65	0.19	0.53	0.41	−0.98
Croatia	0.08	0.06	0.02	−0.40	0.16	0.12	0.03	−0.81	0.24	0.19	0.05	−1.21
Cyprus	0.25	0.26	0.21	0.77	0.51	0.52	0.42	1.53	0.76	0.78	0.63	2.30
Czechia	−0.40	−0.23	−0.23	−0.09	−0.80	−0.45	−0.46	−0.17	−1.20	−0.68	−0.68	−0.25
Denmark	−0.01	0.01	0.02	−0.33	−0.01	0.03	0.05	−0.65	−0.01	0.05	0.09	−1.00
Estonia	0.07	0.08	0.05	0.28	0.15	0.16	0.11	0.54	0.21	0.23	0.15	0.84
Finland	−0.11	−0.01	−0.02	−0.81	−0.22	−0.03	−0.05	−1.63	−0.34	−0.04	−0.07	−2.44
France	−0.03	0.03	0.01	−0.75	−0.06	0.07	0.01	−1.51	−0.08	0.10	0.12	−2.26
Germany	−0.08	−0.02	−0.01	0.16	−0.17	−0.03	−0.02	0.32	−0.25	−0.05	−0.03	0.47
Greece	−0.08	0.04	0.04	0.45	−0.16	0.07	0.07	0.89	−0.24	0.11	0.11	1.34
Hungary	−0.48	−0.23	−0.14	−0.68	−0.96	−0.46	−0.27	−1.36	−1.45	−0.70	−0.41	−2.04
Ireland	0.03	0.03	0.01	−0.28	0.05	0.07	0.02	−0.56	0.08	0.10	0.02	−0.84
Italy	−0.05	0.02	−0.01	−0.56	−0.11	0.03	−0.01	−1.12	−0.16	0.05	−0.02	−1.68
Latvia	0.11	0.07	0.05	−1.53	0.22	0.15	0.10	−3.07	0.32	0.22	0.15	−4.61
Lithuania	−0.39	−0.01	0.05	−3.72	−0.79	−0.03	0.09	−7.40	−1.18	−0.04	0.14	−11.05
Luxembourg	0.05	0.06	0.06	−0.07	0.09	0.12	0.12	−0.14	0.14	0.18	0.18	−0.20
Malta	0.05	0.05	0.03	0.27	0.10	0.10	0.07	0.54	0.16	0.14	0.10	0.81
Netherlands	−0.20	−0.07	−0.05	−0.53	−0.40	−0.14	−0.10	−1.06	−0.60	−0.21	−0.14	−1.59

(continued)

Table 1 (continued)

	Sce 1				Sce 2				Sce 3			
	Prod	GVA	Empl	Emis	Prod	GVA	Empl	Emis	Prod	GVA	Empl	Emis
Norway	−0.11	−0.06	−0.02	−2.07	−0.22	−0.12	−0.03	−4.14	−0.33	−0.18	−0.05	−6.20
Poland	−0.20	−0.12	−0.05	−0.19	−0.41	−0.24	−0.09	−0.38	−0.61	−0.37	−0.14	−0.56
Portugal	0.01	0.14	0.10	−0.57	0.02	0.27	0.21	−1.13	0.03	0.41	0.31	−1.70
Romania	−0.13	−0.03	0.00	−0.52	−0.25	−0.05	0.00	−1.04	−0.38	−0.08	0.01	−1.56
Slovakia	−0.36	−0.12	−0.12	−1.07	−0.71	−0.24	−0.24	−2.13	−1.10	−0.37	−0.34	−3.36
Slovenia	−0.01	0.05	0.04	−2.33	−0.03	0.10	0.09	−4.67	−0.04	0.16	0.13	−7.01
Spain	−0.08	0.01	0.00	−1.26	−0.15	0.03	0.01	−2.53	−0.22	0.06	0.03	−3.80
Sweden	−0.12	−0.06	−0.05	−0.81	−0.24	−0.12	−0.10	−1.61	−0.36	−0.17	−0.15	−2.42
United Kingdom	0.01	0.03	0.03	−0.24	0.02	0.06	0.05	−0.48	0.03	0.08	0.08	−0.72

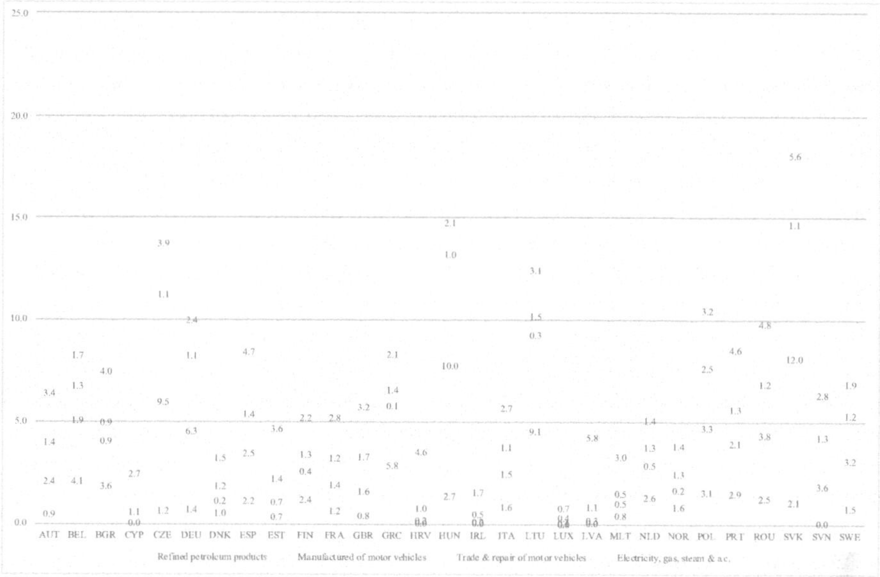

Fig. 1 Relative weight of production sectors on GVA. *Source* Own elaboration based on WIOD [16]

Other countries that show increases in GVA are Bulgaria, Portugal, Estonia, and Latvia. The case of Portugal should be noted, where production increases by 0.02% and GVA, 0.27%. This behavior is due to the different relative weight of the production sectors affected. Thus, the weight of the oil refining sector and manufacture of motor vehicles both represent, 5.0% of the total production while only representing 0.9% of the GVA, which makes the impact of the introduction of EV greater in the case of GVA than in that of production. On the other hand, other countries that see their GVA being reduced are Czechia, Belgium, Poland, and Slovakia.

The impact of EV on employment shows a behavior very similar to the previous variables analyzed (Table 1), with a slight incidence and a disparity in behavior between some countries and others. In this case, for scenario 2, the range of variation ranges between an increase of 0.42%, in the case of Cyprus and a decrease of 0.46% in the case of Czechia. In Cyprus, since it lacks a car industry, the impact on employment in that sector as well as on the Oil Refining branch is practically nil, showing a negative impact on trade and vehicle repair, this being offset by employment growth in most economic sectors. Another country with significant employment growth is Bulgaria with a rate of 0.27%. In this case, the decline in employment in the vehicle manufacturing and commerce sectors is offset by the growth in employment in the other branches of activity.

The opposite case happens in Czechia, where the fall in employment is general in most branches of activity, being especially noteworthy in vehicle manufacturing along with trade and vehicle repair and the rest of the trade. Other countries that have a notable fall in employment are Belgium, Hungary, and Slovakia.

Among the main benefits of introducing EV, in addition to their efficiency and economy, is the reduction of GHG emissions into the atmosphere. Together for the countries studied, and for scenario 2, there is a reduction in emissions that reaches 0.8% of those emitted by all economic sectors. Additionally, although this analysis is focused on the EV impact on the production sectors, it should be taken into account that the benefit of the introduction of EV extends to the household sector, given that a large part of vehicles, passenger cars, belong to households, so the reduction could be between 3 and 4 times higher than that observed for all economic sectors.

As can be seen in Table 1 and Fig. 2, the reduction of GHG emissions as a result of EV penetration is observed in most of the countries coming from the set of economic sectors analyzed. This reduction in GHG emissions is significant in countries such as Lithuania (−7.4%), Slovenia (−4.67%), Norway (−4.14%), Latvia (−3.07%), Spain (−2.53%), and Slovakia (−2.13%). These are countries with a lower emission/production value ratio than the average of the countries studied (see Fig. 3), except in the case of Slovenia and show higher use of renewable energy and/or nuclear power energy to produce electricity (Table 7 of the annex).

On the contrary, there is an increase, although small, in GHG emissions in five countries: Cyprus, Greece, Estonia, Malta, and Germany, with values ranging between 0.32% for Germany and 1.53% for Cyprus. In this case, these are countries with a high ratio of emissions/production value whose cause is the energy mix used for the production of electricity. In the case of Cyprus and Malta, electricity

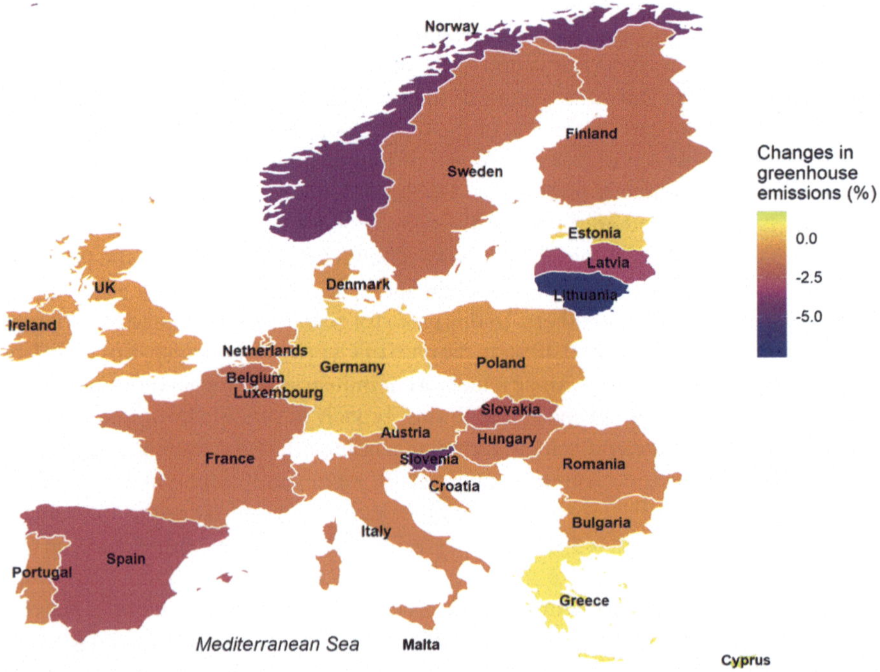

Fig. 2 Changes in greenhouse emissions (%). Scenario 2. *Source* Own elaboration

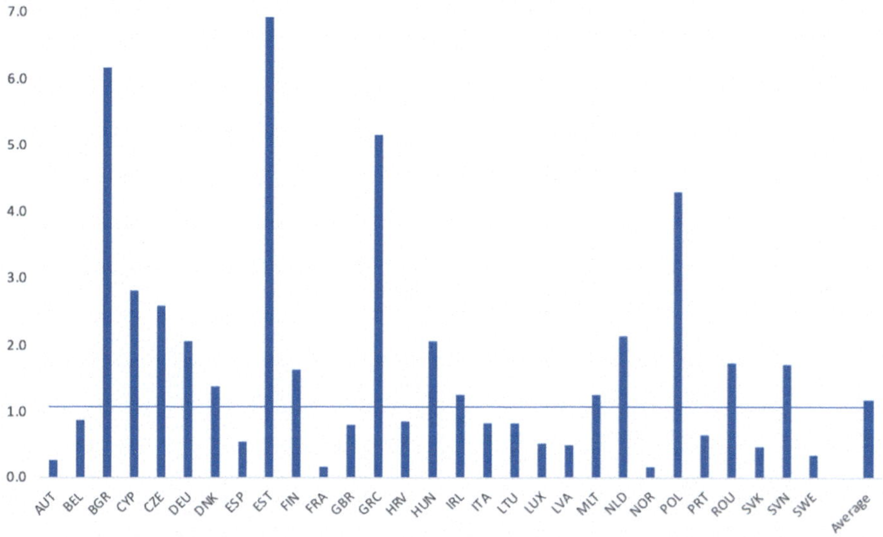

Fig. 3 Emission ratio/Production value of the electricity sector (ktep/million US$). *Source* Own elaboration based on WIOD [16]

production comes almost entirely from oil; in the case of Estonia, the production of electricity comes from the use of shale and oil shale, being the only EU-28 country to use this energy source; and, in the case of Germany and Greece, it is due to the important use of coal as an energy source (see Table 7 of the annex).

If the GHG emissions into the atmosphere are analyzed together with the other three variables studied after the penetration of EV, a different behavior can be observed for the countries studied. Figure 4 shows together the data of the impact of EV on emissions and production for each country. In this figure, four different quadrants can be distinguished. Quadrants I and III include those countries that show a positive correlation between the impact on production and emissions, that is, there is an increase in production accompanied by an increase in emissions or a decrease in production together with a reduction in emissions. In the first group are Cyprus, Estonia, and Malta, and in the second group there are 17 countries among which Lithuania must be highlighted with a significant decrease in both variables; Hungary, Belgium, and Slovakia with greater falls in production than in emissions; and, Norway and Spain, with greater falls in the level of emissions than in production.

In Quadrant II are those countries with a negative behavior of both variables—production has decreased and, nevertheless, the level of emissions has increased—as are the cases of Germany and Greece. In Germany, the automotive sector has a very important weight in the economy and an electricity sector with an energy mix that depends heavily on coal, a highly polluting product. In the case of Greece, the cause must be sought in its oil refining sector with an important weight in its economy and a high use of contaminating energy sources; specifically, the use of solid fuels and oil as sources for obtaining energy exceeds 62% of the total.

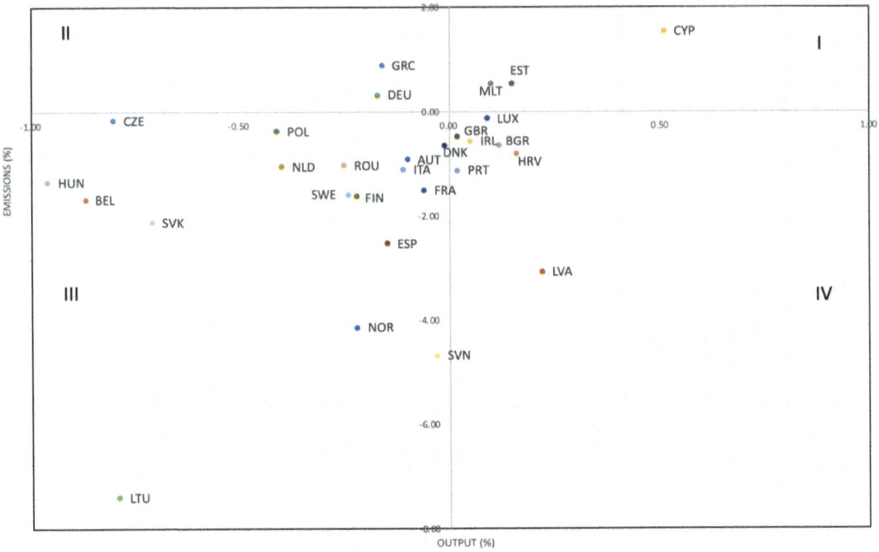

Fig. 4 Changes in emissions and production due to EV penetration in the EU and Norway. *Source* Own elaboration

In Quadrant IV are those countries with a positive behavior in production and emissions, that is, production increases and GHG emissions are reduced. Although the variation is small, the case of Latvia deserves to be highlighted, where the increase in the production of the electricity sector compensates, by far, the minimum decrease observed in the three main sectors affected: oil refining, manufacture of motor vehicles, and trade and repair of vehicles. In the case of the first two, the weight of these sectors in the Latvian economy is very small. On the other hand, the decrease in emissions is mainly due to the land transport sector, its incidence being almost nil in most branches of activity. The increase in emissions in the electricity sector is not high, given that more than half of the electricity is of renewable origin.

The joint analysis of the changes produced in the GHG emissions and the GVA as a result of the penetration of EV shows certain similarities with the previous case, as can be seen in Fig. 5. Most countries are located in the same quadrant. Regarding the differences, it should be noted that there are thirteen countries located in Quadrant IV showing a positive behavior of both variables, that is, they are those that present an increase in GVA and the level of emissions or vice versa. On the contrary, fewer countries now appear in Quadrant III; some countries that appeared in this quadrant in Fig. 4 have now moved to Quadrant IV. In Quadrant II, Germany continues to appear, although with a minimal decrease in GVA and in Quadrant I, Greece is incorporated into the group of Cyprus, Estonia, and Malta.

Finally, the joint analysis of GHG emissions and employment, as seen in Fig. 6, presents results in line with those observed in the previous cases and, more specifically with the relationship between emissions and GVA. The results related to GHG

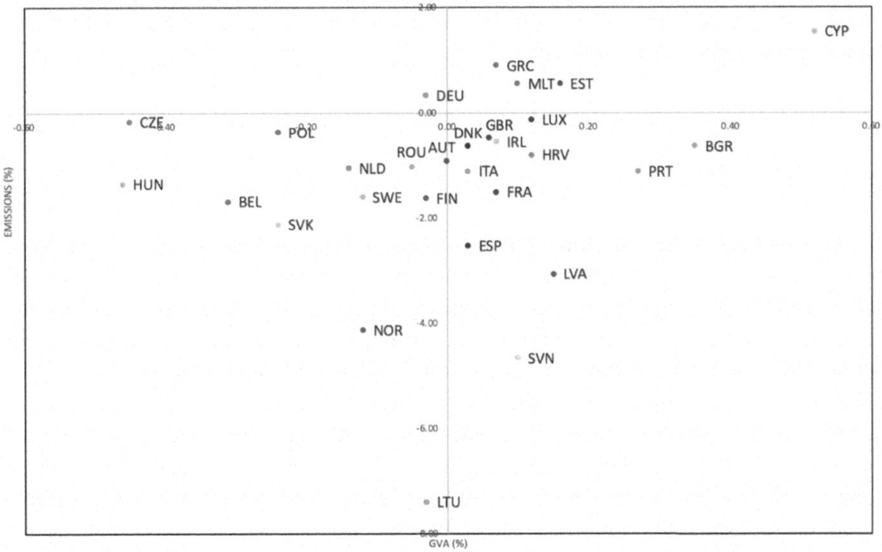

Fig. 5 Changes in emissions and GVA due to EV penetration in the EU and Norway. *Source* Own elaboration

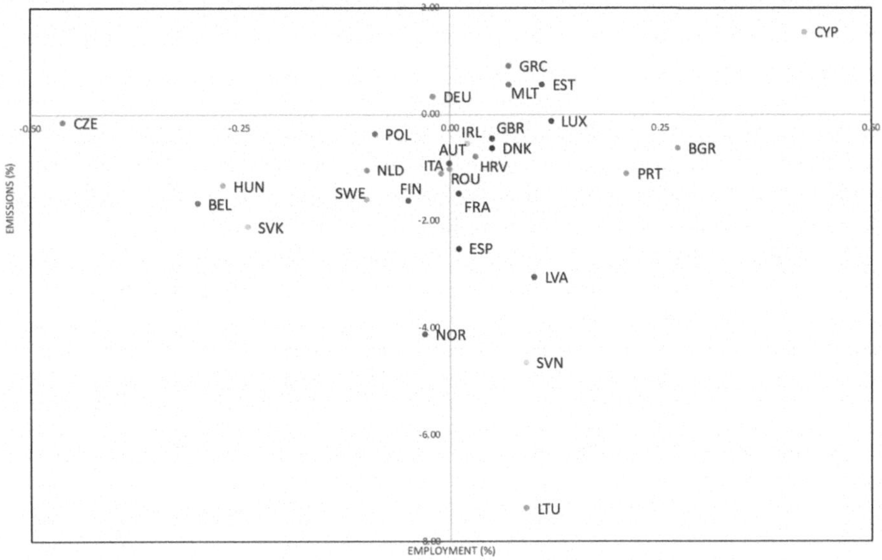

Fig. 6 Changes in emissions and employment due to EV penetration in the EU and Norway. *Source* Own elaboration

emissions/GVA and GHG emissions/employment are very similar, both in terms of changes produced and in their values.

5 Conclusions

The introduction of EV will impact mainly on the following four branches of activity; coking plants and oil refining, manufacture of motor vehicles, trade and repair of motor vehicles, and supply of electrical energy. Through interindustrial relations, the impact is disseminated in the rest of the economic sectors and also in the residential sector, whose purchasing and saving capacity is also modified.

To estimate this impact on production, GVA, employment, and GHG emissions, a conveniently hybridized static IO model has been used to capture all the effects. The central database of the model corresponds to the year 2014, considering three scenarios of penetration of EV of 10, 20, and 30% in the vehicle fleet. The sample of countries analyzed includes the EU 28 and Norway.

For all the variables analyzed, the impacts are, as expected, different between countries. In the case of production, Hungary, the Czech Republic, and Slovakia have negative results while Bulgaria and Croatia are the countries most benefited, only surpassed by small countries such as Cyprus and Malta. In absolute values, the negative impacts outweigh the positive ones. The results obtained show similarities when the variable analyzed is the GVA.

Czechia registers poor results when employment is the variable analyzed, also together with Hungary and Slovakia, although, in this case, Belgium joins the group of countries with the greatest job destruction. Finally, the reduction of GHG emissions could reach 0.8% for the production sectors, increasing to more than triple if the reduction of emissions from the household sector is also included.

The above results should be taken with caution as their validity is in the short term. Longer term assessments would require the use of dynamic modeling.

Acknowledgements The authors are grateful for the financial support received from the Chair of Energy and Environmental Economics sponsored by Red Eléctrica España (REE) at the University of Seville. The first, third, and fourth authors are also grateful to the Spanish Ministerio de Ciencia, Innovación y Universidades for the financial support received from the research project (RTI2018-096725-B-I00). The third and fourth authors acknowledge the support received from the Universidad Autónoma de Chile (Chile).

Annex

See Tables 2, 3, 4, 5, 6 and 7.

Table 2 Percentage of intra-EU and extra-EU exports over total exports by sector[a]

	Refined petroleumn products		Manufacture of motor vehicles		Electricity, gas, steam and air conditioning		Trade and repair of motor vehicles		Total production	
	Intra-EU	Extra-EU	Intra-EU	Extra-EU	Intra-EU	Extra-EU	Intra-EU	Extra-EU	Intra-EU	Extra-EU
Austria	92.2	7.8	71.7	28.3	94.6	5.4	88.0	12.0	69.9	30.1
Belgium	80.9	19.1	73.7	26.3	97.4	2.6	85.1	14.9	72.1	27.9
Bulgaria	7.6	92.4	75.2	24.8	62.7	37.3	67.1	32.9	62.4	37.6
Croatia	0.0	0.0	97.2	2.8	78.7	21.3	76.9	23.1	63.4	36.6
Cyprus					0.0	0.0	54.9	45.1	52.1	47.9
Czechia	100.0	0.0	81.5	18.5	93.8	6.2	90.5	9.5	82.2	17.8
Denmark			72.3	27.7			76.0	24.0	63.6	36.4
Estonia	93.4	6.6	85.4	14.6	99.8	0.2	91.8	8.2	72.3	27.7
Finland			79.7	20.3	83.5	16.5	80.4	19.6	57.3	42.7
France			77.7	22.3			68.8	31.2	60.3	39.7
Germany	65.7	34.3	44.6	55.4	88.4	11.6	66.5	33.5	57.7	42.3
Greece	19.8	80.2	50.8	49.2	60.6	39.4	77.1	22.9	48.3	51.7
Hungary	84.2	15.8	86.5	13.5	27.5	72.5	83.4	16.6	80.0	20.0
Ireland			90.2	9.8					55.0	45.0
Italy	27.9	72.1	63.2	36.8	53.9	46.1	74.7	25.3	54.9	45.1
Latvia	0.0	0.0	71.1	28.9	99.6	0.4	92.4	7.6	68.5	31.5
Lithuania			69.9	30.1	82.0	18.0	47.8	52.2	54.8	45.2
Luxembourg							98.3	1.7	82.6	17.4
Malta									49.9	50.1
Netherlands							83.1	16.9	75.9	24.1

(continued)

Table 2 (continued)

	Refined petroleumn products		Manufacture of motor vehicles		Electricity, gas, steam and air conditioning		Trade and repair of motor vehicles		Total production	
	Intra-EU	Extra-EU	Intra-EU	Extra-EU	Intra-EU	Extra-EU	Intra-EU	Extra-EU	Intra-EU	Extra-EU
Norway	91.6	8.4	82.5	17.5	100.0	0.0			81.8	18.2
Poland	73.2	26.8	87.5	12.5	97.1	2.9	76.6	23.4	77.4	22.6
Portugal	0.0	0.0	96.0	4.0	87.3	12.7	75.0	25.0	70.8	29.2
Romania	26.5	73.5	73.1	26.9	61.2	38.8	91.1	8.9	71.1	28.9
Slovakia			73.0	27.0			94.2	5.8	84.4	15.6
Slovenia			89.3	10.7	98.5	1.5	79.9	20.1	75.3	24.7
Spain	35.0	65.0	79.1	20.9	21.6	78.4	91.5	8.5	64.0	36.0
Sweden			62.0	38.0			78.4	21.6	58.5	41.5
United Kingdom	57.6	42.4	40.8	59.2	80.3	19.7	58.8	41.2	47.9	52.1

[a]For the cells that appear without data for the sector considered, the data of the total exports of each country have been taken

Source Eurostat [8] and Statistics Norway [13]

Table 3 Reduction of final demand in the manufacture of motor vehicles branch

	(1)	Sce 1	Sce 2	Sce 3
EV penetration (%)		10	20	30
Demand reduction in parts and accessories (%)		8	16	24
	(1)	Decrease in final demand in the manufacture of motor vehicles branch (%)		
Austria	32.04	2.56	5.13	7.69
Belgium	23.10	1.85	3.70	5.54
Bulgaria	95.93	7.67	15.35	23.02
Croatia	78.52	6.28	12.56	18.85
Cyprus	30.56	2.44	4.89	7.33
Czechia	51.98	4.16	8.32	12.47
Denmark[a]	28.87	2.31	4.62	6.93
Estonia[a]	28.87	2.31	4.62	6.93
Finland	9.38	0.75	1.50	2.25
France	28.01	2.24	4.48	6.72
Germany	21.83	1.75	3.49	5.24
Greece	41.62	3.33	6.66	9.99
Hungary	42.89	3.43	6.86	10.29
Ireland	85.74	6.86	13.72	20.58
Italy	39.50	3.16	6.32	9.48
Latvia	68.82	5.51	11.01	16.52
Lithuania	44.85	3.59	7.18	10.77
Luxembourg[a]	28.87	2.31	4.62	6.93
Malta[a]	28.87	2.31	4.62	6.93
Netherlands	17.74	1.42	2.84	4.26
Norway	63.84	5.11	10.21	15.32
Poland	57.06	4.56	9.13	13.69
Portugal	59.91	4.79	9.59	14.38
Romania	66.01	5.28	10.56	15.84
Slovakia	37.47	3.00	6.00	8.99
Slovenia	46.81	3.74	7.49	11.23
Spain	29.99	2.40	4.80	7.20
Sweden	20.22	1.62	3.24	4.85
United Kingdom	19.21	1.54	3.07	4.61

(1)% manufacture of parts and accessories branch over the total of manufacture of motor vehicles branch

[a]The EU-28 data has been taken for the countries indicated

Source Eurostat [5]

Table 4 Decrease in final demand in the Trade and Repair of motor vehicles branch

	(1)	Sce 1	Sce 2	Sce 3
EV penetration (%)		10	20	30
Demand reduction in parts and accessories (%)		8	16	24
	(1)	Decrease in final demand in the manufacture of motor vehicles branch (%) (%)		
Austria	41.37	3.31	6.62	9.93
Belgium	45.65	3.65	7.30	10.96
Bulgaria	60.35	4.83	9.66	14.49
Croatia	48.98	3.92	7.84	11.76
Cyprus	64.03	5.12	10.24	15.37
Czechia	54.47	4.36	8.72	13.07
Denmark	49.35	3.95	7.90	11.84
Estonia	50.87	4.07	8.14	12.21
Finland	51.57	4.13	8.25	12.38
France	48.40	3.87	7.74	11.61
Germany	43.87	3.51	7.02	10.53
Greece	49.44	3.96	7.91	11.87
Hungary	41.85	3.35	6.70	10.04
Ireland	53.45	4.28	8.55	12.83
Italy	58.05	4.64	9.29	13.93
Latvia	63.88	5.11	10.22	15.33
Lithuania	60.99	4.88	9.76	14.64
Luxembourg	24.83	1.99	3.97	5.96
Malta[a]	47.18	3.77	6.75	10.12
Netherlands	23.10	1.85	3.70	5.55
Norway	61.59	4.93	9.85	14.78
Poland	53.19	4.26	8.51	12.77
Portugal	45.19	3.61	7.23	10.84
Romania	65.64	5.25	10.50	15.75
Slovakia	42.18	3.37	6.75	10.12
Slovenia	49.07	3.93	7.85	11.78
Spain	58.12	4.65	9.30	13.95
Sweden	47.21	3.78	7.55	11.33
United Kingdom	45.57	3.65	7.29	10.94

(1)% manufacture of parts and accessories branch over the total of manufacture of motor vehicles branch

[a]The EU-28 data has been taken for the countries indicated

Source Eurostat [6]

Table 5 Increase in final demand in the electricity, gas, and water branch. Abbreviations

	AUT	BEL	BGR	HRV	CYP	CZE	DNK
Average annual route per vehicle (km)	15,000	15,000	15,000	15,000	15,000	15,000	15,000
Energy consumption per electric vehicle (watt hours/km)	250	250	250	250	250	250	250
Annual energy consumption per vehicle (kWh)	3,750	3,750	3,750	3,750	3,750	3,750	3,750
Vehicle fleet (cars) year 2014 (units)	4,694,921	5,555,499	3,013,863	1,474,000	478,492	4,833,386	2,329,578
Electric vehicles (10%)	469,492	555,550	301,386	147,400	47,849	483,339	232,958
Total consumption of electric vehicles (GWh)	1,760.6	2,083.3	1,130.2	552.8	179.4	1,812.5	873.6
National electricity consumption year 2014 (GWh)	62,557	82,585	28,887	15,180	3,972	55,086	31,569
Increase in electricity consumption (%)	2.81	2.52	3.91	3.64	4.52	3.29	2.77
Generation, transport and distribution (% prod s/branch)	67.34	87.86	85.30	77.63	100.00	69.29	47.83
Increase in demand for electricity (%). Sce 1	1.90	2.22	3.34	2.83	4.52	2.28	1.32
	EST	FIN	FRA	DEU[a]	GRC	HUN	IRL[b]
Average annual route per vehicle (km)	15,000	15,000	15,000	15,000	15,000	15,000	15,000
Energy consumption per electric vehicle (watt hours/km)	250	250	250	250	250	250	250
Annual energy consumption per vehicle (kWh)	3,750	3,750	3,750	3,750	3,750	3,750	3,750
Vehicle fleet (cars) year 2014 (units)	652,950	3,172,735	32,531,000	44,403,000	5,110,873	3,107,695	2,018,310
Electric vehicles (10%)	65,295	317,274	3,253,100	4,440,300	511,087	310,770	201,831
Total consumption of electric vehicles (GWh)	244.9	1,189.8	12,199.1	16,651.1	1,916.6	1,165.4	756.9
National electricity consumption year 2014 (GWh)	7,417	80,429	429,217	525,904	51,185	35,899	24,313
Increase in electricity consumption (%)	3.30	1.48	2.84	3.17	3.74	3.25	3.11
Generation, transport and distribution (% prod s/branch)	76.98	91.00	66.04	85.65	98.18	66.96	82.01
Increase in demand for electricity (%). Sce 1	2.54	1.35	1.88	2.71	3.68	2.17	2.55
	ITA	LVA	LTU	LUX[b]	MLT[b]	NLD	NOR
Average annual route per vehicle (km)	15,000	15,000	15,000	15,000	15,000	15,000	15,000

(continued)

Table 5 (continued)

	ITA	LVA	LTU	LUX[b]	MLT[b]	NLD	NOR
Energy consumption per electric vehicle (watt hours/km)	250	250	250	250	250	250	250
Annual energy consumption per vehicle (kWh)	3,750	3,750	3,750	3,750	3,750	3,750	3,750
Vehicle fleet (cars) year 2014 (units)	37,080,753	657,799	1,205,668	372,827	265,950	7,979,083	2,555,000
Electric vehicles (10%)	3,708,075	65,780	120,567	37,283	26,595	797,908	255,500
Total consumption of electric vehicles (GWh)	13,905.3	246.7	452.1	139.8	99.7	2,992.2	958.1
National electricity consumption year 2014 (GWh)	291,085	6,582	10,009	6,182	2,005	107,291	115,676
Increase in electricity consumption (%)	4.78	3.75	4.52	2.26	4.97	2.79	0.83
Generation, transport and distribution (% prod s/branch)	80.18	71.83	58.93	82.01	82.01	99.93	96.34
Increase in demand for electricity (%). Sce 1	3.83	2.69	2.66	1.85	4.08	2.79	0.80

	POL	PRT	ROU	SVK	SVN	ESP	SWE	GBR
Average annual route per vehicle (km)	15,000	15,000	15,000	15,000	15,000	15,000	15,000	15,000
Energy consumption per electric vehicle (watt hours/km)	250	250	250	250	250	250	250	250
Annual energy consumption per vehicle (kWh)	3,750	3,750	3,750	3,750	3,750	3,750	3,750	3,750
Vehicle fleet (cars) year 2014 (units)	20,003,863	4,699,645	4,908,000	1,949,100	1,068,360	22,029,512	4,584,711	30,250,294
Electric vehicles (10%)	2,000,386	469,965	490,800	194,910	106,836	2,202,951	458,471	3,025,029
Total consumption of electric vehicles (GWh)	7,501.4	1,762.4	1,840.5	730.9	400.6	8,261.1	1,719.3	11,343.9
National electricity consumption year 2014 (GWh)	136,307	46,139	45,808	25,083	12,559	233,306	125,041	310,807
Increase in electricity consumption (%)	5.50	3.82	4.02	2.91	3.19	3.54	1.37	3.65
Generation, transport and distribution (% prod s/branch)	58.75	92.40	71.82	67.60	91.39	86.19	76.51	75.88
Increase in demand for electricity (%). Sce 1	3.23%	3.53	2.89	1.97	2.92	3.05	1.05	2.77

[a]Calculation made on the GVA
[b]Data from UE-28

Source Eurostat [5, 9, 10]

Table 6 Gross household saving rate. Year 2014

	%		%
Austria	12.83	Latvia	−1.88
Belgium	12.49	Lithuania	0.16
Bulgaria	−4.51	Luxembourg	20.60
Croatia*	10.89	Malta*	10.89
Cyprus	−6.20	Netherlands	16.34
Czechia	11.77	Norway	13.89
Denmark	4.21	Poland	2.26
Estonia	10.67	Portugal	5.19
Finland	7.18	Romania*	10.89
France	14.25	Slovakia	7.21
Germany	16.82	Slovenia	12.44
Greece*	10.89	Spain	9.25
Hungary	12.94	Sweden	18.81
Ireland	8.03	United Kingdom	8.60
Italy	11.30	EU-28	10.89

Source Eurostat [7]

Table 7 Share of production of electricity and derived heat by type of fuel (%). Year 2014. Abbreviations

	Solid fossil fuels	Manufact. gases	Peat & peat products	Oil shale and oil sands	Natural gas	Oil and petroleum products	Renewables and biofuels	Non renewable waste	Nuclear heat
AUT	4.5	3.0	0.0	0.0	8.3	0.9	82.2	1.1	0.0
BEL	3.1	3.0	0.0	0.0	26.7	0.3	18.5	1.7	46.6
BGR	44.9	0.0	0.0	0.0	4.5	0.4	16.7	0.0	33.4
HRV	17.5	0.0	0.0	0.0	7.4	1.0	74.2	0.0	0.0
CYP	0.0	0.0	0.0	0.0	0.0	92.7	7.3	0.0	0.0
CZE	47.3	3.3	0.0	0.0	2.1	0.1	11.9	0.1	35.2
DEN	34.4	0.0	0.0	0.0	6.5	1.0	55.9	2.2	0.0
EST	0.0	4.1	0.4	82.8	0.6	0.3	11.2	0.6	0.0
FIN	11.7	0.7	5.0	0.0	8.1	0.3	38.7	0.6	34.8
FRA	2.0	0.5	0.0	0.0	2.3	1.1	17.4	0.4	76.4
DEU	43.9	1.7	0.0	0.0	10.0	0.9	26.9	1.2	15.5
GRC	51.0	0.0	0.0	0.0	13.4	11.0	24.4	0.2	0.0
HUN	20.4	0.4	0.0	0.0	14.4	0.3	10.7	0.4	53.3
IRL	15.2	0.0	9.6	0.0	48.4	1.0	25.6	0.3	0.0
ITA	15.6	1.1	0.0	0.0	33.5	5.1	43.8	0.9	0.0
LVA	0.0	0.0	0.0	0.0	45.5	0.0	54.5	0.0	0.0
LTU	0.0	0.0	0.0	0.0	42.1	3.9	52.9	1.0	0.0
LUX	0.0	0.0	0.0	0.0	48.9	0.0	49.2	1.9	0.0
MLT	0.0	0.0	0.0	0.0	0.0	96.7	3.3	0.0	0.0
NLD	28.6	2.8	0.0	0.0	49.9	1.8	11.3	1.6	4.0

(continued)

Table 7 (continued)

	Solid fossil fuels	Manufact. gases	Peat & peat products	Oil shale and oil sands	Natural gas	Oil and petroleum products	Renewables and biofuels	Non renewable waste	Nuclear heat
NOR	0.0	0.1	0.0	0.0	1.8	0.0	97.9	0.2	0.0
POL	81.5	1.3	0.0	0.0	3.4	1.0	12.8	0.0	0.0
PRT	22.6	0.0	0.0	0.0	12.9	2.6	61.4	0.5	0.0
ROU	27.0	0.1	0.0	0.0	12.3	0.7	42.0	0.0	17.8
SLK	10.5	1.8	0.0	0.0	5.9	1.1	23.8	0.1	56.8
SLN	21.6	0.0	0.0	0.0	2.1	0.2	39.5	0.0	36.5
ESP	15.7	0.5	0.0	0.0	17.0	5.1	40.9	0.2	20.6
SWE	0.2	0.3	0.1	0.0	0.3	0.2	55.9	0.7	42.2
GBR	29.6	0.4	0.0	0.0	29.8	0.6	19.9	0.7	18.9

References

1. Cansino JM, Yñiguez R (2018) Promoting electro mobility in Spain. Public measures and main data (2007–2012). Trans Res Part D: Transp Environ 59:325–345
2. European Environment Agency (2019) Transporte. https://www.eea.europa.eu/es/themes/transport/intro
3. European Parliament (2013) Propuesta de resolución sobre CARS 2020: Hacia una industria europea del automóvil fuerte, competitiva y sostenible (2013/2062(INI). http://www.europarl.europa.eu/sides/getDoc.do?pubRef=-//EP//TEXT+REPORT+A7-2013-0391+0+DOC+XML+V0//ES#top
4. Eurostat (2019a) Air emissions accounts by NACE Rev. 2 activity. www.ec.europa.es/eurosat/)
5. Eurostat (2019b) Annual detailed enterprise for industry (NACE Rev. 2). www.ec.europa.es/eurosat/
6. Eurostat (2019c) Annual detailed enterprise for trade (NACE Rev. 2). www.ec.europa.es/eurosat/
7. Eurostat (2019d) Household saving rate. www.ec.europa.es/eurosat/
8. Eurostat (2019e) International trade in goods, Trade by NACE Rev. 2. www.ec.europa.es/eurosat/
9. Eurostat (2019f) Road transport equipment-stock of vehicles. www.ec.europa.es/eurosat/
10. Eurostat (2019g) Supply, transformation and consumption of electricity. www.ec.europa.es/eurosat/
11. Eurostat (2019h) Production of electricity and derived heat by type of fuel. www.ec.europa.es/eurosat/
12. Leontief W (1966) Input-output economics, 2ª edn. Oxford University Press, New York
13. Statistics Norway (2019) External economy, External trade in goods. www.ssb.no
14. Timmer MP, Dietzenbacher E, Los B, Stehrer R, de Vries GJ (2015) An illustrated user guide to the world input-output database: the case global automotive production. Rev Int Econ 23:575–605
15. World Input-Output Database (2018) Socio economic accounts release 2016. www.wiod.org/database/seas16/
16. World Input-Output Database (2016) National input-output tables. Released. www.wiod.org/database/niots16/

The Interest of Mineral Raw Materials in the Development of Electric Vehicles

Luis de la Torre Palacios, Eloy Álvarez Pelegry, and Jose Antonio Espí Rodríguez

Abstract This paper examines the outlook for the supply and demand of mineral raw materials, as related to strong growth in the introduction of electric vehicles (EVs) in the near future. Given the nature of the topic to be addressed, this analysis will be global in scope. It will also focus on batteries, as the fundamental element differentiating EVs from internal combustion vehicles. An analysis of the readiness of the supply side to respond to a major increase in demand for the various raw materials involved shows very substantial differences between the different supply chains, including those of substances classified as critical. No less important are the geopolitical consequences that might threaten some aspects of the market.

1 Introduction to Electric Vehicles

Before assessing and discussing possible future problems with the availability or price of mineral raw materials as a result of the growing penetration of electric vehicles (EVs), we first need to specify which types of EV we are talking about. Secondly, we need to identify in which part of the vehicle the mineral raw materials are located. And thirdly, it is essential to estimate the number of EVs that will be operating in a given timescale.

In Spain, the National Policy Framework (*Marco de Acción Nacional*) defines EVs as vehicles that are fully or partially driven by an electric motor using chemical energy stored in one or more batteries charged from an external power source. The

This paper is an updated and revised version of the article 'Protagonismo de las materias primas minerales en el desarrollo del vehículo eléctrico' by Luis de la Torre Palacios, Eloy Álvarez Pelegry and Jose Antonio Espí Rodríguez, published in Economía Industrial No. 411.

L. de la Torre Palacios · J. A. E. Rodríguez
Polytechnic University of Madrid, Madrid, Spain

E. Á. Pelegry (✉)
Royal Academy of Engineering of Spain, Madrid, Spain
e-mail: ipelegry@gmail.com

A. Arcos-Vargas (ed.), *The Role of the Electric Vehicle in the Energy Transition*, Green Energy and Technology, https://doi.org/10.1007/978-3-030-50633-9_8

NPF's definition includes Battery Electric Vehicles (BEVs), plug-in hybrid electric vehicles (PHEVs) and range-extended electric vehicles (REEVs).

This study narrows the scope of BEVs and PHEVs. It does not cover non-rechargeable conventional hybrids, fuel cell electric vehicles[1] (FCEVs)—which use hydrogen—or REEVs, due to their low rate of market penetration in recent years.

It is also important to distinguish between different usage segments. The first of these—and the most important in terms of number of units currently on the road and anticipated growth—is passenger light-duty vehicles (PLDVs) and light commercial vehicles (LCVs). Two- and three-wheeled vehicles, such as electric bicycles, are also important. The final segment comprises buses and trucks.

By 2017, the world's total stock of EVs numbered three million units, with China accounting for 40% of the total. Global sales in 2017 exceeded one million units. In addition, 100,000 buses and 30 million two-wheeled electric vehicles were sold that year; again the majority in both cases were sold in China [9].

2 An Introduction to EV Components

One way of viewing the future impact of EV penetration on materials is to compare an electric vehicle with one with an internal combustion engine. This is the approach taken by UBS in its 2018 study (UBS 2018), which compares a Chevrolet Bolt to a VW Golf.

The total weight of the raw materials in the two vehicles comes to around 1600 kg in the case of the Bolt and 1300 kg in the case of the Golf. The Bolt is 22% heavier than the Golf, primarily because of the battery weight. The Bolt contains 70% more aluminium; 80% more copper; 75% less steel; 60% less iron and 100% less platinum-group metals (PGMs).

At the same time, the Chevrolet Bolt has 140 kg of 'active' materials in its batteries (nickel, cobalt, lithium, manganese and graphite) and one kilogram of rare earth elements in the electric motor, particularly neodymium and dysprosium.

In the Chevy Bolt, according to the same source, steel accounts for around 39% of the total weight, iron 2%, aluminium 9%, copper 5%, rubber 1%, graphite 3%, manganese 2%, cobalt 1%, nickel 2%, lithium 0.6%, while rare earth elements and other components make up a considerable 31%.

The UBS report uses these figures to project a world where 100% of vehicles are electric. In such a scenario, there would be an increase in global demand of 2,511% for lithium, 1.2% for cobalt, 26% for graphite, 11% for nickel, 100% for rare earth elements, 21% for copper, 135% for manganese and 12% for aluminium. Demand for steel would fall by 1% and for PGMs by 53%. In economic terms, the value of the semiconductors in the Bolt is estimated at about USD 580, 6–10 times as much as the Golf.

This approach, which seeks to assess the impact of transport electrification on the demand for mineral raw materials, is founded on the basic assumption that demand for raw materials is driven by battery materials.

Any analysis of the demand for raw materials needs to take into account not only increased battery production, but also demand related to power generation and storage, power grids, charging infrastructures and, naturally, demand from the EVs themselves [8].

2.1 Batteries

There is broad consensus today that in the medium-term future at least, ion-lithium batteries will continue to form the bedrock of EV battery development.

This prediction is partly based on the development of ion-lithium batteries in consumer electronics, which has led to the accumulation of extensive experience and an important reduction in unit costs.

EV batteries contain a number of chemicals which are significant because of their influence on the demand for materials. For example, the cathode or positive electrode contains lithium-nickel-manganese-cobalt (NMC); lithium-nickel-cobalt-aluminium oxide (NCA); lithium-manganese oxide (LMO) and lithium ferrophosphate (LFP). In most current designs, the anode is made of graphite, but lithium-titanate (LTO) is also used, especially for heavy vehicles Warner [16] in IEA [9].

The IEA [9] has previously highlighted the importance of developments in transport electrification on demand for materials. The agency points to three major changes, namely: an increase in the use of copper, the rare earth elements contained in electric motors and scarce metals.

It is important to examine the intensity of what the European Union defines as strategic—or in some cases critical—materials (critical raw materials, or CRMs are defined as combining raw materials of high importance to the EU economy and of high risk associated with their supply). These are lithium (Li), nickel (Ni), cobalt (Co) and manganese (Mn).

The IEA identifies the intensity of different battery chemicals in kg/kWh (see Table 1).

Table 1 Intensity of critical metals in the chemistry of the main battery types (kg/kWh)

	Li	Ni	Co	Mn
NCA	0.1	0.67	0.13	0.00
NMC 111	0.15	0.4	0.40	0.37
NMC 433	0.14	0.47	0.35	0.35
NMC 532	0.14	0.59	0.23	0.35
NMC 622	0.13	0.61	0.19	0.20
NMC 811	0.11	0.75	0.09	0.09
LFP	0.1	–	–	–

Source ANL (2018b). *BatPaC: A Lithium-Ion Battery Performance and Cost Model for Electric-Drive Vehicles* in IEA [9]

2.2 *Demand for Electric Vehicles and Batteries*

2.2.1 Demand for Batteries and Related Materials

There will be two essential consequences of the large-scale roll-out of EVs: an increased demand for electricity and an increased demand for batteries. In the case of power demand, consumption hypotheses range between 18 and 27 kWh/100 km. The IEA [9] estimates 20–27 and 18 kWh/100 km [1].

We also need to consider the figure for yearly mileage figure, where the estimates of the two sources cited in Table 2 range from 8,500 to 20,000 km per year.

Using forecast vehicle figures for 2030, with the highest and lowest estimated consumption per kilometre and annual mileage, gives us a very wide range of estimates for global electricity demand in 2030, ranging from 168 to 594 TWh for 110 million vehicles, to 348–1,231 TWh for 228 million vehicles in the EVI30@30s 'ambition' target. (The EV30@30 campaign, launched in 2017, sets a target for members of the Electric Vehicles Initiative (EVI) of reaching a 30% sales share for EVs by 2030).

We also need to consider power consumption by two- and three-wheeled vehicles, as well as trucks and buses. The IEA [9] gives a figure of 404 TWh and 928 TWh for 2030 in the NPS and EVI scenarios cited above.

Any increase in the number of EVs and electricity demand will be linked to an increase in battery capacity and production.

By 2030, the battery range of EVs is expected to increase, translating into an increase in battery capacity of around 70–80 kWh (as compared to the current figure

Table 2 Estimated demand for lithium, nickel, cobalt and manganese by battery model shown in Table 1 (in thousand tonnes)

	Intensity in tonnes/GWh				Prediction based on 775 GWh in 2030				Prediction based on 2000 GWh in 2030			
Type	Li	Ni	Co	Mn	Li	Ni	Co	Mn	Li	Ni	Co	Mn
NCA	100	670	130	0	77.50	519.25	116.25	0	200	1,340	300	0
NMC 111	150	400	400	370	116.25	310.00	310.00	286.75	300	800	800	740
NMC 433	140	470	350	350	108.50	364.25	271.25	271.25	280	940	700	700
NMC 532	140	590	230	350	108.50	457.25	178.25	271.25	280	1,180	460	700
NMC 622	130	610	190	200	100.75	472.75	147.25	155.00	260	1,220	380	400
NMC 811	110	750	90	90	85.25	581.25	69.75	69.75	220	1,500	180	180
LFP	100	–	–	–	77.50	–	–	–	200	–	–	–

Source Prepared by the authors

of 20 kWh in China and 60 kWh in the United States according to the IEA [9]; this would mean a rise in annual battery capacity from 68 GWh in 2017 to 775–2250 GWh in 2030 depending on the scenario.

Other sources cite figures of 450 TWh of power demand in 2030, with 8,000 GWh of batteries in operation in light vehicles worldwide [11]. Based on EV fleet and sales figures, this would entail a production capacity of roughly 2,000 GWh in that year.

In order not to exhaust readers with any further digressions, we shall base our estimate for demand for battery materials in the two scenarios on assumed figures of 775 GWh and 2000 GWh.

Using these figures, Table 2 shows estimated demand for lithium, nickel, cobalt, and manganese in thousand tonnes for 2030, depending on the type of battery.

The IEA [9] gives a central 2030 estimate of relative content of battery chemicals of 50% NMC 811, 40% NMC 622 and 10% NCA. Based on this estimate, in the New Policy Scenario (NPS), demand would come to 101,000 tonnes for cobalt and 91,000 tonnes for lithium; naturally, these figures are far higher in the EV30@30 scenario, with figures of around 291,000 and 263,000 tonnes, respectively.

Taking sales of 10 million and 30 million EV units in 2025 and 2030 respectively, demand for nickel would come to 299,000 and 985,000 tonnes and for cobalt 80,000 and 259,000 tonnes respectively [8]. Glencore also estimates increases in copper demand in the two timeframes of 1 and 2 million tonnes, respectively.

Using the figures in the Table 2, we can position these estimates in relation to vehicle demand and battery chemicals. For example, changes in the chemical make-up of the cathode will have a greater effect on demand for Co and Li, primarily.

2.3 *Supply of Metals Used in EV Power Accumulators*

2.3.1 The Materials that Make Up the Electric Vehicle

On the supply side, this section addresses the problem of the availability of the raw materials used in the fundamental parts of EVs.

We divide these materials into two groups: those that make up the structure of the vehicle, and those that are part of the electrical accumulation systems and electric motors. The former involve demand-flexible production chains, while in the case of the latter, major efforts will be required to adapt the industries involved in their production. They will also need continuous transformation, in order to respond adequately to greater-than-trend demand.

The second group includes some minerals and metals that are relatively scarce (cobalt and rare earths) and others that are less so (graphite, lithium, nickel and manganese), but for which a rise in demand is anticipated (although the exact size of this increase is disputed). We shall further discuss the source and supply problems of the natural materials used in electric batteries.

This section address general features related to their emergence on the market, from their initial emergence to their decline or disappearance, as well as the specific characteristics of their supply chains.

2.3.2 Graphite

Part played by graphite in EVs. This is the most common component in many electric car batteries, particularly in lithium-ion batteries.

Conditions of natural reserves. Natural graphite is the product of metamorphic recrystallization of organic matter contained in rocks. When this process occurs in layers of coal, or in rocks that contain liquid hydrocarbons, deposits of graphite are generated. However, deposits of this ore can also be volcanic or hydrothermal in origin. Natural graphite is relatively abundant in nature and consists mostly (80–90%) of carbon, with inorganic impurities of different kinds. In the concentration process, these impurities are eliminated using selective flotation systems or chemical treatments. The properties and composition of the graphite are determined by its geological location. It is hardly surprising, therefore, that a priori graphite resources seem immense.

There are three different types of natural graphite produced in different types of mineral deposits:

a. Flake graphite. This is the least common form of graphite. Its costs about 4 times more than amorphous graphite and is used in many traditional applications. In addition, it is highly sought-after for graphite applications such as the anode material of lithium-ion batteries.
b. Amorphous graphite. This is the most abundant form of graphite. It has a comparatively low (70–80%) carbon content. It has no visible crystallinity and is the least pure of the three types. It is not of sufficient quality for use in most electrical accumulation applications.
c. High crystalline graphite (crystalline vein). This form of graphite is extracted only in Sri Lanka. It has a carbon content of 90–99%. Its scarcity and high cost limit its viability for most applications.

Additionally, there is also synthetic graphite, manufactured using high-temperature treatment of amorphous carbon materials. The raw material used in the process is calcined petroleum coke and coal tar, making it very expensive to produce—up to 10 times the cost of natural graphite.

Concentration of production. In 2018, China was the world's largest source, producing 630,000 tonnes of ore, according to the US Geological Survey, February 2019 (Table 3), which calculates the country's share at 68% of total world extraction and 35% of consumption. Despite China's absolute dominance of the graphite market, however, this position is not expected to continue indefinitely. Brazil is the world's second-largest source, producing 95,000 tonnes of graphite. In third place comes India, at 35,000 tonnes. Today, there is a clear concentration of production

Table 3 World graphite production

Country	Graphite 2018 production t
China	630,000
Brazil	95,000
Canada	40,000
India	35,000
Mozambizue	20,000
Other	110,000
World total	930,000

Source Own elaboration based on US Geological Survey [15]

(quantified in Table 3). Above all, the Chinese predominance should be a cause for concern.

Possible substitutes. New uses of technology in fuel cells, batteries and applications, such as high strength lightweight composites may substantially increase the global demand for graphite, as there are currently no substitutes.

New resources. Despite its great abundance in nature, graphite (particularly its higher-value qualities) is under-researched. The figure of 800 million tonnes inferred by the USGS [14] is considerably lower than other estimates of close to one billion tonnes. Large-scale flake varieties are in great demand for applications in quality products, including the manufacture of graphene. In the short term, prospecting for graphite in as-yet unstudied geological environments is expected to bear fruit. In addition, artificial graphite may always be considered as an alternative in high-end products, although its production cost currently appears prohibitive.

2.3.3 Lithium

Part played by lithium in EVs. The newly created lithium-ion batteries are formed by a lithium salt electrolyte and graphite electrodes and cobalt oxide. The use of new materials such as lithium has made it possible to achieve high specific energies, high efficiency, elimination of the memory effect and a lack of maintenance. In addition, they have twice the energy density of nickel-cadmium batteries and are around one-third smaller. However, they also have disadvantages, the main one being their high production cost, although this is gradually being reduced. They are fragile and can explode if overheated and must be stored very carefully.

Conditions of natural reserves. Lithium is a relatively rare element, although due to its abundance in the earth's crust, it is listed as the 27th most common element. Despite being found in many rocks and some brines, its concentration is usually very low. High-concentration lithium brines come from both geothermal waters and surface leaching from volcanic ashes, clays or other rocks. The brines may be geothermal (long exploited), from oil fields (with enormous possibilities and relatively well-studied) and from heterolytic clays (a very abundant and promising

Table 4 Global lithium production

Country	Lithium 2018 production in t (Li content)
Australia	51,000
Chile	16,000
China	8,000
Argentine	6,200
Zimbabwe	1,600
Other	1,900
World total	83,100

Source Own elaboration based on US Geological Survey [15]

source for the future). Approximately half of all lithium currently produced comes from conventional hard rock deposits, while the other half comes from the extraction of lithium dissolved in brines.

The US Geological Survey [15] estimates global reserves of lithium (mineral resources that can be economically exploited) from solid lithium minerals, brines and minerals in clays (heterolytic) at 55 million tonnes.

Concentration of production. The leading producers of lithium (see Table 4) are Australia, Chile, China, Argentina and Zimbabwe. Lithium in very large quantities has also been identified in the brines of Bolivia, China and Israel. It is estimated that China and Europe are the world's largest consumers of lithium, accounting for 29% and 28% of the total, respectively. Until now, lithium production sources and demand have been relatively well-balanced. However, over the last year supply has exceeded demand, causing a—possibly transitory—surplus. Nevertheless, a deficit in the supply of lithium is coming, with new countries joining the demand.

In reality, there are no major differences in the production potentials of supplier countries. Moreover, in the near future, sources of lithium supply will be further diversified, with the inclusion of lithium from brines from oil wells, anomalous clays in lithium and others. In addition, there are only small differences in production costs between hard ore mined using conventional methods and lithium extracted by pumping from brines

New resources. As demand for lithium increases, new players are expected to join the supply side. Strategic investors are already taking positions in lithium source types not previously exploited, such as heterolytic clays and brines from oil fields, both of which would have a greater yield than pegmatites with spodumene and other lithium minerals.

2.3.4 Cobalt

Part played by cobalt in EVs. In a lithium-ion battery, the different cathode chemicals have an impact on the demand for the component raw materials. For example, by weight, LCO (lithium-cobalt oxide) batteries contain only 7% cobalt and 60%

lithium, while an NMC (nickel-manganese-cobalt) battery has approximately 7% lithium, 20% nickel, 19% manganese and 22% cobalt in relation to total weight. An indicative figure (depending on battery type) shows that each battery contains approximately 15 kg of cobalt chemicals. Lately there has been a growing awareness that different cathode technologies in lithium-ion batteries may allow progress to be made towards reducing cobalt and replacing it with nickel.

Conditions of natural reserves. Both in its free and combined state, metal cobalt resembles iron and nickel. It is widely distributed in nature and makes up approximately 0.001% of the total igneous rocks in the earth's crust, as compared to 0.02% for nickel. Cobalt and its alloys are resistant to wear and corrosion, even at very high temperatures. It is important to bear in mind that lithium-ion batteries actually contain no cobalt metal. Rather, they contain cobalt chemicals, with cobalt sulphate being one of the raw materials preferred by cathode manufacturers.

In nature, cobalt is part of a series of mineral deposits, whose main features are as follows:

- There are no clear concentrations. Moreover, although taken as a whole there are extensive deposits, no more than five types are of economic interest.
- Sediment Hosted (SH) copper deposits specifically included in strata or sedimentary episodes, are the model of greatest generic interest, accounting for over half of the world's production.
- It is therefore unsurprising that cobalt production is geographically concentrated, with a concentration in the African interior. In other cases, production is associated with other metals and depends on whether they are economical to mine.
- Marine nodules, although not constituting a resource for immediate use, will play an important role in the future.

Concentration of production. Cobalt deposits are found worldwide but are most prevalent in the African copper belt (the Democratic Republic of the Congo and Zambia), with more than 64% of the world's cobalt production concentrated in a single country (Table 5).

In 2016, approximately 60% of the cobalt extracted was a by-product of copper, 38% a by-product of nickel and the remaining 2% came from primary cobalt mines.

Table 5 Global cobalt production

Country	Cobalt 2018 production in t
DRC	90,000
Russia	5,900
Cuba	4,900
Australia	4,700
Philippines	4,600
Other	29,900
World total	140,000

Source Own elaboration based on US Geological Survey [15]

Thus, changes in the global production of copper and nickel are the main determinants of changes in cobalt production.

New resources. Cobalt from abyssal marine nodules can only be considered as an alternative to traditional deposits. Cobalt reserves are estimated at 7.2 million tonnes with total resources of 25 million tonnes. However, about 120 million tonnes of cobalt are in the form of manganese nodules in abyssal bottoms in the Atlantic, Indian and Pacific oceans. Nonetheless, any real exploitation of these resources still faces legal and environmental barriers, as well as technological difficulties.

2.3.5 Nickel

Part played by nickel in EVs. The original lithium-ion batteries introduced by Sony in 1991 used a lithium-cobalt or LCO cathode powder, which was approximately 60% cobalt by weight. Although LCO has remained the chemical of choice for personal electronic products for almost 30 years, it has never been seen as an enabling chemical for electric vehicles, since it is scarce and expensive and LCO cells have a spectacular safety record.

In 1999, two nickel-rich cathode chemical compounds were introduced. The first of these is nickel-cobalt-manganese, or NCM/NMC, which uses equal proportions of nickel, cobalt and manganese to reduce the cobalt content from 60 to 20%. In addition, nickel-cobalt-aluminium (NCA) chemistry mainly uses nickel with small amounts of cobalt and aluminium to reduce the cobalt content from 60 to 9%. Since 1999, battery manufacturers have continued their efforts to reduce cobalt content, but the pace of progress has been very modest.

Conditions of the natural stock. Nickel is a metallic element of natural origin, lustrous and silvery white. It is the fifth most common element on earth and appears extensively in the earth's crust. However, most of the nickel is found in the centre of the earth and is therefore inaccessible. The key features of nickel metal are high melting point, resists corrosion and oxidation, very ductile, easily alloyed, magnetic at room temperature, can be deposited by electroplating and has catalytic properties.

Due to these characteristics, nickel is extensively used in over 300,000 products for consumer, industrial, military, transportation, aerospace, marine and architectural applications. Its greatest use (about 65%) is in alloys, especially with chromium and other metals to produce stainless and heat-resistant steels. In many of these applications, there is no substitute for nickel that does not reduce performance or increase costs.

Nickel mineral resources consist of primary sulphide minerals (45%) with an average Ni content of 0.58%, and laterite ores (55%) with an average Ni content of 1.32%. Only 42% of world production comes from laterite type minerals, while the remaining 58% comes from sulphide minerals. It is estimated that 72% of the world's mineral resources are included in lateritic minerals, while 28% of all global mineral resources are in sulphide minerals.

Concentration of production. The five main nickel producing countries in 2018, according to the latest figures from the US Geological Survey (Table 6) are: Indonesia,

Table 6 Global nickel production

Country	Nickel 2018 production in t
Indonesia	560,000
Philippines	340,000
Russia	210,000
New Caledonia	210,000
Australia	170,000
Other	810,000
World total	2,300,000

Source Own elaboration based on US Geological Survey [15]

the Philippines, Russia, New Caledonia and Australia.

According to the Herfindahl–Hirschman Index, there is no concentration in production, although some countries—such as Indonesia, Philippines, Russia and New Caledonia—have retained a prominent position for many years.

New resources. Over the past two decades, lithium-ion battery manufacturers have eagerly sought advanced cathode formulations that would partially replace expensive cobalt through the use of much cheaper nickel. In general, increasing the nickel content in a cathode formulation improves the energy density of the battery, but reduces stability, meaning that there is a trade-off between cost and safety.

Nickel has been widely explored throughout the world for many years, due to the value of its concentrates. As a result, good information is available on its production potential and economic outlook. Identified land-based resources averaging 1% nickel or greater contain at least 130 million tonnes of nickel, with about 60% in laterites and 40% in sulphide deposits [14]. Nickel, together with cobalt and copper, forms part of the composition of marine nodules, which in the future could be an alternative to current resources.

2.3.6 Manganese

Part played by manganese in EVs. Manganese is an essential element for modern industry. Its main use is in the manufacture of steel. Although the amount of manganese consumed to make a tonne of steel is small (0.6–0.9%) it is an irreplaceable component in its production.

Conditions of the natural stock. Manganese is the twelfth most abundant in the earth's crust, accounting—although estimates vary—for about 0.15%. The highest quality manganese minerals contain 40–45% manganese. The predominant processes in the formation of the world's main deposits take place in marine environments.

Concentration of production. There should in principle be no global shortage of manganese mineral resources, albeit a number of strictly economic factors limit production. For example, widespread use in steel for construction has led to intense demand (around 18 million tonnes) (Table 7) and it is in this large-scale production

Table 7 Global manganese production

Country	Manganese 2018 production in Kt
South Africa	5,500
Gabon	2,300
Australia	3,100
China	1,800
Brazil	1,200
Other	4,100
World total	18,000

Source Own elaboration based on US Geological Survey [15]

that the metal might become scarce. Production concentration is a consequence of the need to achieve large-output mining projects.

According to the US Geological Survey [5], the Kalahari manganese district in South Africa contains 70% of the world's identified resources and about 25% of its reserves.

2.4 *Considerations on Production Concentration: The Herfindahl–Hirschman Index (I)*

The Herfindahl–Hirschman Index (I) is used in economics to measure market economic concentration. A high I score denotes a highly concentrated and uncompetitive market. The index is calculated by squaring the market share owned by each country and totalling these amounts. Thus, a perfect monopoly, in the case of producing countries, would give an I score of 10,000.

To homogenize the results as much as possible, the same source of information has been used to calculate all five substances, viz. the statistics provided by the USGS for 2017 (Table 8).

Table 9 shows the Herfindahl–Hirschman Index for different minerals. Rare earths occupy a prominent first place, although there has been an important decline in the score compared to previous years, when it constituted an almost perfect monopoly. An important concentration can also be seen in the production of graphite, reflecting the strong Chinese presence on the market. However, in this case, the situation is not as dramatic as with rare earths, since all analysts recognize that there are opportunities in other countries. It is also striking to note the similar position in the scores for lithium and cobalt, which are both undergoing a dramatic process of adjustment to future demand, which will leave cobalt in a more isolated position, since the geographical distribution of new production opportunities will not result in any change in its score. Manganese and nickel are in the last position, as explained by the maturity of their production—the result of extensive demand from the metallurgical industry.

Table 8 Principal producing countries of mineral raw materials, in percentages

Rare earth producing countries	% world 2017 production	Graphite producing countries	% world 2017 production	Cobalt producing countries	% world 2017 production
China	83	China	65	Australia	41
Australia	11	India	14	Chile	34
Russia	2	Brazil	7	Argentina	16
Thailand	<1	Turkey	3	China	5
Malaysia	<1	Mexico	2	Zimbabwe	2
Other	2		9		2
Thailand	<1	Turkey	3	China	5
Malaysia	<1	Mexico	2	Zimbabwe	2
Other	2		9		2
Cobalt producing countries	% world 2017 production	Manganese producing countries	% world 2017 production	Nickel producing countries	% world 2017 production
DRC	54	South Africa	29	Philippines	22
Canada	6	China	19	Russia	11
China	6	Australia	15	Canada	11
Russia	5	Gabon	12	Australia	9
Australia	4	Brazil	7	New Caledonia	9
Other	25		25		38

Source Prepared by the authors, based on production data by country, USGS 2017

Table 9 Herfindahl–Hirschman index (I) of the five substances analysed in EVs

	Herfindahl–Hirschman index (I)
Rare Earths	7,016
Natural graphite	4,492
Lithium	3,122
Cobalt	3,119
Manganese	1,688
Nickel	976

Source Prepared by the authors, based on production data by country, USGS 2017

2.5 *The Formation of Supply Chains and the Problems of Secure Supply*

2.5.1 Concepts

The structure of the material supply chain needed by an industrial sector is often quite complex and very specific to that industry. If we consider the initial part of an industrial product's life cycle, it starts with the supplier's relationship with the natural environment in which the raw materials are obtained—in this case, the minerals. The problems begin when a mineral good is produced in excessive volumes and with temporary demand. Knowledge of the natural stock of raw materials is always complex and full of uncertainties and it is therefore crucial to begin new supplies to cater to sudden variations in demand.

It is well understood that metal concentrations need time to be discovered in nature and put into production. Project maturity periods of over 12 years are not unusual and a figure of around 8 years is very common. Depletions in mineral concentrations should result in their efficient substitution and, when this occurs, technology must come to their aid. To facilitate an understanding of these phenomena, a very simple diagram of the value chain (production cycle) is shown in Fig. 1.

The 'value chain' of the mineral raw material production process begins in Stage 1 of the life cycle of each metal. This phase involves exploration for new resources using selectivity criteria based on the desired qualities. Selection of these resources is related to market demand or, in any case, economies of production. Stage 2 involves mining the mineral raw material. The dimensions and grades are related, both by the qualities and quantities of the deposit and by the demand. Selectivity in exploitation is determined by grade limits, which are based on demand and production costs.

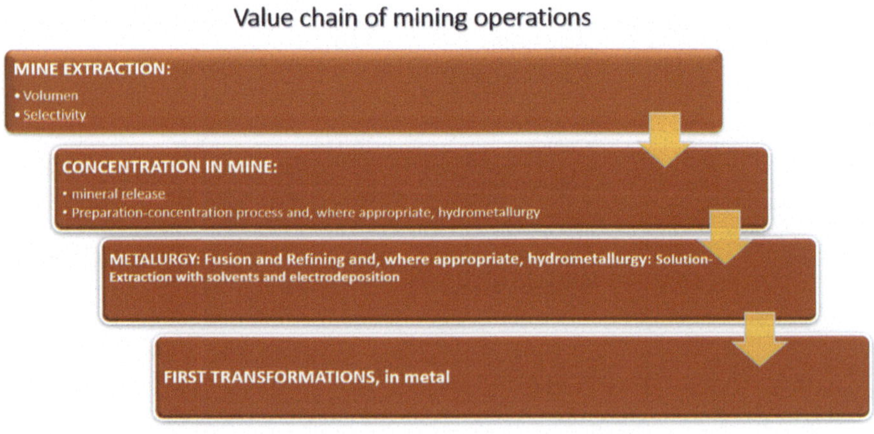

Fig. 1 Value chain in most common mining operations. Authors design

In Stage 3, the extracted ore is concentrated and prepared at the place of mining; this process often requires a greater technological effort and operating costs. In this operation, the grains of ore must be released for concentrating, in order to produce products of the highest possible grade. They are then transported from the exploitation site to the metallurgy or refining process.

In Stage 4, a complex industrial process is required to extract the metal from the containing mineral. This operation is carried out in different locations away from the mine and consumes large amounts of energy. In some cases, it may be considered to form part of Stage 3, since an impure metal can be produced in the hydrometallurgical process associated with this stage.

Finally, Stage 5 includes a number of secondary processing operations which, in the case of metals, turn the raw material into ingots or plates. In some cases, these products are used in further industrial processing, as part of the final chain in making more complex components. In other cases, as with lithium or graphite, these operations do not produce metals but entail a chemical transformation to create the final component.

2.5.2 Application to the Components of Electric Batteries

Table 10 shows the metals and minerals used in electric vehicles whose value (supply) chains face uncertainty due to a foreseeable extraordinary demand. Two levels of criticality are identified in the table in light and dark grey. The table also shows the most frequent metal concentrations at each stage of the life and supply cycle.

Table 10 Possible criticality in the different phases of the EV metal value (supply) chain due to critical demand for EV batteries and metal concentrations in each phase

	Graphite	Lithium	Cobalt	Nickel	Manganese	Neodymium
Stage 1	80–90% C	>1% Li_2O –500 ppm Li_2O	>0.1% Co	>0.4% Ni	>4% Mn	>0.1% Rare Earths
Stage 2	80–90% C	>1% Li_2O –500 ppm Li_2O	>0.1% Co	>0.4% Ni	>4% Mn	>0.1% Rare Earths
Stage 3	>90% C	>5% Li_2O	>2% Co	5–12% Ni	>70% Mn	>40% Rare Earths
Stage 4	>95% C	Li Carbonate	Variable as chemical products	Variable as chemical products	Variable as chemical products	>85% Neodymium hydroxide
Stage 5	>95% C	Li Carbonate	Very high	Very high	Very high	>95% Neodymium hydroxide

Source Own elaboration

2.5.3 Explanation and Comments

Graphite

In the case of graphite, the supply chain is not directed towards the production of a pure component, but rather its physical preparation for incorporation into the manufacturing stages of the battery components. For this reason, the quality of the deposit (Stage 1) is fundamental, since higher mineral qualities allow for more demanding final products, as is the case with new batteries. Graphite is very abundant in nature, with some agencies estimating up to 900 million tonnes of recognized resources; however, the best quality deposits are very scarce.

Lithium

Like graphite, lithium is not used in elemental or isolated form, but chemically in the final stages of the supply chain. Lithium can come from host solid minerals (spodumene, petalite and others), in relatively low concentrations (generally no more than 4% Li_2O) in Stage 1 or 2, while in the concentration process (Stage 3) it can reach about 10% Li_2O. Stage 4 involves chemical processing, to obtain the lithium carbonate product that is marketed for the first phases of incorporation into lithium-ion batteries.

For some years, underground or surface brines (the Andean salt flats) have offered an alternative source to solid minerals, and today, the global lithium supply is distributed fairly equally between the two sources. The operating advantage of this method is that the lithium is dissolved and Stage 1 or 2 of the value-supply chain almost overlaps with Stage 3 (concentration). The rest of the cycle is largely similar.

Lithium metal is very abundant in nature and Stage 1 of the supply chain is constantly being enriched with the discovery of new potential occurrences (new sources of supply) and any scarcity of the mineral is therefore highly unlikely. Nonetheless, from an economic perspective, when working with very low metal concentrations, not all sources and qualities are admissible for Stage 3. These conditions may vary according to price, but battery economics may also suffer due to the current high prices of this metal. Stage 5 also requires proper analysis, because around 60% of manufacture of the final product is concentrated in China.

Cobalt

One of the most notable characteristics of cobalt production is the concentration of producers in Stage 1. In addition, production is shared with copper and nickel. Stage 3 (the concentration process) is very similar to that of other metals and does not entail any specific problems, except for the difficulties of expanding the size of existing facilities to adapt production to demand. There are no large cobalt deposits in the world (Stage 1) and the alternative exploitation of marine nodules is still at an experimental phase. China also has a strong role in Stage 4 of the cobalt supply chain.

Nickel

The nickel used in Stage 1 comes from two very clearly defined sources. Approximately half of all production comes from sulphide deposits with a relatively high concentration of nickel. The remainder is found in oxidized form, with very low metal grades. In the future, however, oxidized nickel deposits will be the chief sources of supply, as it is here that there is greatest potential for an increase in production. Stage 3 is quite dissimilar in the two cases with a difference in electrical consumption per unit of metal produced. In the metallurgical stage (Stage 4), there is an even greater difference, and work is ongoing to reduce the electrical consumption of oxidized and lateritic minerals.

Manganese

Manganese is a relatively abundant metal in nature and forms mineral deposits of all categories (Stage 1). Its extraction and processing do not involve different processes to other minerals, apart from some cases in which hydrometallurgical or chemical processes are used (Stages 3 and 4). The final product of the mining and beginning of the metallurgical process (Stage 5), for supply to the accumulator industry, may entail rigidities due to a strong demand for chemical products, a phenomenon which is reflected in current prices.

2.5.4 Consequences for Secure Supply of Mineral Raw Materials for Electric Vehicle Batteries

Based on current supply conditions and medium-term forecasts for EV mineral components (or their constituent metals) and the circumstances that may affect secure supply of the electric battery industry, the following considerations should be taken into account for each of the components:

Graphite

- Existence of an adequate value-supply chain.

In today's world, graphite is considered a key strategic material in the economics of green technology, which includes advances in energy storage, electric vehicles, photovoltaic energy and electronics. Graphite is also the source of graphene. As the green tech economy grows over the next decade, demand for graphite is expected to outstrip supply. It is thought that, in the EV market alone, the estimated demand for 2020 would require more than today's total production. With the demand for large-scale graphite production growing, it is estimated that 25 new graphite projects will be needed to meet the world's needs in 2021.

Emerging markets such as India and China, where the pace of industrialization has far exceeded world averages, have led to a slowdown in the supply of graphite for domestic consumption. China still controls more than 65% of global graphite production, although recently there has been a fall in output, and some older and smaller mines have been closed due to environmental violations. China is also consuming more graphite itself, withdrawing a small amount from international market in order to export the finished products.

At the same time, it is believed that the world's recognized graphite resources may be in excess of 900 million tonnes. In other words, the gap between supply and demand cannot be blamed on the research effort, but rather on a lack of adaptation for many causes, with China at the centre of the controversy. In the short term, however, adapting the supply chain to the avalanche of demand for quality products motivated by new energy technologies should not be an insurmountable problem, especially in view of the possibility of manufacturing artificial graphite.

– *Political factors.*

Asia-Pacific countries, driven mainly by China and India, constitute the fastest-growing market for graphite. Factors such as low labour costs and natural graphite resources are leading to sustainable market growth (especially in China), even in low demand conditions. With a solid position on the graphite market, China is expected to see sustained growth, backed by its overseas investments; however, it also recently began to protect its internal needs and to control its exports.

Lithium

– Existence of an adequate value-supply chain.

According to Goldman Sachs, global demand for lithium increased by 39% in 2018. Given that the metal is used not only in batteries for electric vehicles and mobile devices, but also for the manufacture of lubricating greases and other uses, one must ask whether there will be enough lithium to go around in the coming decades. With 40 million tonnes of currently recognized resources, the availability of this element is not currently a factor that limits large-scale production of electric cars. What might slow the rate of battery manufacture, however, are bottlenecks in the lithium distribution chain.

There are also some question marks over the potential for rapid adaptation of world lithium production to strong demand. There is concern that supply is slowing down due to the complexity of building evaporation ponds in Andean regions of South America which, as well as matters of cost, also pose problems related to water supply and other environmental issues.

Demand rose in 2018 by over 27% and is expected to grow at over 20% in 2019. According to Ricardo Ramos (CEO SQM) beyond 2019, the prospect for growth in demand continues to look extremely healthy, leading us to believe that the landmark of a 1 million ton per year lithium market may be reached sooner than originally anticipated [10].

Many analysts believe that there is enough lithium metal in the Earth's crust to support the manufacture of electric vehicles in the required volumes, based solely on the needs of lithium batteries. Although assessments are far from being precise, it is thought that there are very abundant available resources of lithium.

Twenty-six battery plants are expected to begin production or expand capacity by 2022. In 2014, there were only three battery mega factories in process. These plants have a combined planned capacity of 344 GWh. To put this in perspective, total demand for lithium-ion batteries in 2017 is estimated at 100 GWh, but the industry needs to expand. Indeed, it is estimated that demand for batteries could rise to between 775 GWh and 2000 GWh by 2030 (see estimates in Table 2). This would involve lithium demand levels of between 77,500 and 300,000 tonnes.

Adapting the lithium supply chain to foreseeable demand is likely to involve market dislocation, based on EV production figures from 2050 on (although in this report we base our analysis and assumptions on predictions for 2030), due to the immense production effort required to meet demand. This might appear catastrophic, but the real scenario will not be quite as dramatic. We can readily envisage the first stage of adaptation with an increase in production to 2030, motivated by current high prices and interest by investors in new projects. Subsequent adjustments in price will come as a result of reduction in battery consumption and increased efficiency, as has always been the case.

- Political factors.

China is the world's largest consumer of lithium, due to its rapid economic development, large population and growing demand for electric vehicles—driven by the search for solutions to air pollution problems, particularly in certain cities. China's lithium resource depends heavily on imports, with 70% of the spodumene concentrate imported only from Australia. Projected growth in electric vehicle sales will ensure that the country remains dependent on lithium imports and even in China, some commentators are already predicting problems with security of supply.

2019 is a good example of volatility in the lithium market [10]. Despite a double-digit growth in demand, prices continued their downtrend. As a result, lithium miners around the world reacted to the challenging market conditions by scaling back production and cutting costs. China is a key driver because they are the main consumers of lithium raw materials. The change in the electric vehicle subsidy regime in China, combined with a relative weakness in the China economy, has impacted lithium raw material demand, in turn impacting pricing in unexpected ways.

Lithium prices fell consistently throughout 2019, resulting in lithium off-takers and strategic groups being more cautious with their approach. This was further exacerbated by a number of unexpected negative macro factors, including the magnitude of US–China trade tensions and China revoking EV subsidies.

Cobalt

– The existence of an adequate value and supply chain.

In 2020, consumption in batteries is predicted to account for 59% of all cobalt demand, reflecting a 58% increase in demand for batteries compared to 2016 levels [7], which is expected to result from increased demand for electric vehicles.

The rechargeable battery segment has become the largest- and fastest-growing end use of cobalt. Around 97% of global cobalt production is a secondary by-product of copper and nickel extraction, leaving cobalt supplies exposed to fluctuations in the copper and nickel market. If the demand for copper drops, cobalt production could fall with it. The suspension of Glencore's production in its copper and cobalt mines Katanga and Mopani in the Democratic Republic of the Congo (DRC) and Zambia in 2015 is a good example.

So, although a large proportion of the existing cobalt supply remains uncertain and it is difficult to obtain new capacity due to the shortage of primary cobalt resources, the risk to the lithium-ion battery supply chain remains, and prices are likely to continue rising.

– Political factors.

China controls most of the world's refined cobalt and depends on the Democratic Republic of the Congo for more than 90% of its cobalt supply. In the words of specialist Rawles [13] 'The important thing is to realize that China produced 80% of the world's cobalt chemicals and that much of its raw material comes from concentrates from the Democratic Republic of the Congo'. Any change in the producing country can have a real impact on the prices of cobalt chemicals.

Nickel

– Existence of an adequate supply chain.

Nickel differs from the other raw materials discussed in having a well-consolidated value/supply chain. Nickel is present in stainless steels that make up an important part of the flow of industrial material, especially in products and applications of quality steels. It is important to consider that total world production of nickel metal comes to

2.3 million tonnes (USGS 2018) and the foreseeable quantities required in EVs have not yet been incorporated. In addition, as already mentioned, the excessive increase in cobalt prices has already aroused interest in replacing it with another very similar metal—for example, nickel. Even in nature, the two metals are found together in some types of mineral deposits.

Nonetheless, the roll-out of electric cars could lead to demand for this metal rising by 2030 to anywhere between 310,000 and over 1.5 million tonnes—very significant figures compared to current production levels.

- Political factors.

Judging by the values shown in the HHI index in Table 9, nickel is free from the dangers of cartelization, as production is quite diversified. In addition, somewhat unusually among the minerals discussed here, the Chinese presence is not a cause for concern, given its low production levels.

Manganese

- Existence of an adequate supply chain.

In the early stages of the cycle, the mining industry does not differentiate by final destination of the product (steel industry and manufacture of chemical products). However, both its concentrates and, in certain cases, the precipitates from leaching in the manganese plant, are specifically oriented towards the final chemicals. Products derived from them may be used as elements in electric accumulators.

For all these reasons, and also at a global level, supply chains may be able to cope with major increases in demand, especially when price differences arising from that situation are not important.

- Political factors.

Manganese is a critical metal for the steel industry of some important countries, such as the United States. The fact is that, although more moderate than other strategic metals, manganese production is relatively concentrated. However, we are dealing here with the supply of manganese as chemicals for the manufacture of electric batteries, not the construction steel industry, and for this industry, although the Chinese presence is very important, it seems unlikely that pressure can be exerted on the supply of this metal.

3 Conclusions

An adequate and secure supply of mineral raw materials is crucial for the current and future EV industry, if it is to successfully complete the progressive and intense roll-out of electric vehicles. Based on our understanding of the problems that are expected to arise in the coming years, the viability of efficient and safe supply chains will be of key importance in addressing growth in demand for mineral raw materials to meet the foreseeable global increase in electric vehicles.

Meeting the need for energy storage has been a much sought-after goal for many decades. However, it is now—with the backing of more resolute social opinion and an automotive industry that is beginning to accept the new challenge—that technology must be used comprehensively to achieve that goal competitively, in a way that meets the needs of consumers.

In the problem analysed here of catering to increases in demand for raw materials and metals for EV batteries, three key points need to be highlighted:

1. *Uncertainty over demand, with two fundamental aspects not fully defined.* The first area of uncertainty is the quantification of demand, with lower and higher values varying greatly from one forecast to another. The second aspect involves the slope or gradient of growth, reflecting the key importance of the speed of change. This could also lead to a lack of synchrony between technical restrictions and their solution, as well as a sharp variation in prices caused by the threat of a presumed shortage.
   As regards technical restrictions, of the five phases of the supply chain, the critical phases identified for each material are: graphite, Phase 1 (exploration); lithium, Phase 3 (preparation and concentration); cobalt, practically all phases; nickel, Phase 4 (metallurgy of concentrates); manganese, preparation of concentrates and metallurgy; and neodymium, the first and last phases.
   As far as prices are concerned, minerals such as lithium—where there is uncertainty about the possible appearance of short-term bottlenecks in coming years—may be more clearly affected. The price of cobalt, too, has almost quadrupled in three years due to the threat of a market shortage. Thus, prices will be the economic signals for the development of new productions, with physical and logistical capacity requiring periods of adjustment during which prices will be high.
   However, in the case of raw materials that are clearly abundant in nature, the supply chain can be relied upon to adapt gradually in the medium term. In the case of resources such as cobalt, for which there is no medium-term guarantee of supply, past experience suggests that the most likely solution will lie with technological advances in the search for substitutes.
   The search for substitute goods—such as a higher proportion of nickel to reduce the cobalt content in batteries—will also affect value-supply chains. Therefore, while uncertainties with regard to demand estimates will more clearly affect lithium and cobalt, it is important to note that prices will also play a role as a variable.

2. *Time horizons of forecasts.* While demand will clearly see very considerable growth to 2030, estimates of the scale of that increase vary greatly and it is practically impossible to make any reasonably reliable estimates for longer-term horizons. This is an extremely important consideration, given the rate of response and transformation of the mining industry. Discovering new mineral resources and developing mining projects inevitably takes time and the accuracy of the models depends precisely on this response time. For some metals, especially cobalt, there is no reasonable response to an acceleration in demand of the order predicted after the 2020s.
3. *The danger of geopolitical exploitation of weaknesses in the supply chain.* China holds a dominant position at different points in the supply chain of almost all scarce mineral raw materials required for the electric vehicles of the future, a factor which must at least be a cause for concern. It is very striking how the Asian giant has managed to take a preferential position, both in the production of raw materials and in the primary processing of low production and almost always strategic minerals. This is the case for all raw materials currently classed as strategic and sometimes critical (CRM). It should not be forgotten that China is the country with the highest future demand for electric vehicles. Moreover, the concentration of production in very few hands (the highest Herfindahl–Hirschman scores are for rare earths, graphite, lithium and cobalt) and areas of political or social instability (cobalt) is another factor of insecurity often recognized by the markets themselves.

Note

1. In 2017, there were 7,200 units of this type of vehicle on the road. Of these, slightly more than half were in the United States, 2,300 in Japan and 1,200 in Europe, mainly Germany. For more information on the basic techniques and use of hydrogen in transport, see Alvarez Pelegry and Menendez Sanchez [1].

References

1. Alvarez Pelegry E, Menendez Sanchez J (2017) *Energias alternativas para el transporte de pasajeros. El caso de la CAPV: análisis y recomendaciones para un transporte limpio y sostenible.* Cuadernos Orkestra
2. Andrew M (2015) Flake graphite market trends and pricing patterns. Industrial Minerals INATA indmin.com/graphiteanalysis
3. Barrera P (2019) Lithium forecast 2020: execs expect recovery, bet on long-term fundamentals. Investing news. https://investingnews.com/daily/resource-investing/battery-metals-investing/lithium-investing/lithium-forecast/?mqsc=E4106167
4. Bloomberg (2018) New energy outlook 2018. https://about.bnef.com/new-energy-outlook
5. Cannon WF, Kimball BE, Corathers LA (2017) Manganese. Professional Paper 1802-L. U.S. Geological survey. https://pubs.usgs.gov/pp/1802l.pdf
6. CRU (2018) CRU International Limited. Cobalt analysis. https://crugroup.com/analysis/technology-metals

7. Darton Commodities Limited (2018) dartoncommodities.co.uk
8. Glencore (2018) The EV revolution and its impacts on raw materials. https://www.iea.org/media/Workshops/2018/Session3Glencore.pdf
9. IEA (2018) Global EV outlook 2018: towards cross-modal electrification. www.iea.org
10. Investing News (2019) The battery metals market in 2019. https://investingnews.com/free-report-online/battery-metals-market-stocks/
11. IRENA (2018) Electric vehicles technology brief. www.irena.org
12. Kay A (2018) 10 top manganese-producing countries. Investing Nes Netor. https://investingnews.com/daily/resource-investing/critical.metals-investing/manganese-investing
13. Rawles C (2018) Cobalt critical to lithium-ion battery chemical mix. Investing News Network
14. USGS (2018) Mineral commodity summaries. Graphite (Natural). https://minerals.usgs.gov/minerals/pubs/commodity/graphite
15. U.S. Geological Survey (2019) Mineral commodity summaries. https://prd-wret.s3-us-west-2.amazonaws.com/assets/palladium/production/s3fs-public/atoms/files/mcs-2019-graph.pdf
16. Warner J (2015) The handbook of lithium-ion battery pack design. Amsterdam, Oxford Woltham. Elsevier Science

Alternative Energies in Transport in the Context of Energy Transitions

Eloy Álvarez Pelegry and Macarena Larrea Basterra

Abstract Alternative energies in transport are essential to advance toward a low-carbon economy. Besides the transport sector is responsible for a high share of greenhouse gas and pollutant emissions. Nevertheless, in the energy transition processes, little has been done concerning transport. However, during the last years, European institutions have developed quite an abundant number of rules to foster different energy alternatives for road transport. There are different energy alternatives, with distinct economic and environmental characteristics. As alternative energies infrastructure seems to be relevant to achieve the penetration of alternative energies, the last section of the document shows the present situation of energy alternatives vehicles and infrastructure development for some European countries.

1 Purpose and Scope

In November 2019, the Communication from the Commission known as "The European Green Deal" established the objective to achieve climate neutrality by 2050, increasing the ambition both for 2030 and 2050.

In that communication among the essential elements, to both passenger and freight transport, there is the first European Climate Law (planned to be released by March 2020) and the revision of the Energy Taxation Directive (to harmonize fuel pricing across the EU), considering alternative fuels as a necessity to meet the sector's growth and the role of modal cooperation (between the different transport modes) [21].

Regarding alternative fuels, in 2014, the European Commission released the Directive 2014/94/EU on the deployment of alternative fuels infrastructure. "*Alternative fuels*" *means fuels or power sources which serve, at least partly, as a*

E. Álvarez Pelegry (✉)
Royal Academy of Engineering of Spain, Madrid, Spain
e-mail: ipelegry@gmail.com

M. Larrea Basterra
Orkestra-Fundación Deusto, Universidad de Deusto, Bilbao, Spain

Deusto Business School, Universidad de Deusto, Bilbao, Spain

A. Arcos-Vargas (ed.), *The Role of the Electric Vehicle in the Energy Transition*, Green Energy and Technology, https://doi.org/10.1007/978-3-030-50633-9_9

substitute for fossil oil sources in the energy supply to transport and which have the potential to contribute to its decarbonization and enhance the environmental performance of the transport sector [23]. They include, inter alia: electricity, hydrogen, biofuels as defined in point (i) of Article 2 of Directive 2009/28/EC, synthetic and paraffinic fuels, natural gas, including biomethane, in gaseous form (compressed natural gas (CNG)) and liquefied form (liquefied natural gas (LNG)), and liquefied petroleum gas (LPG).

The Directive 2014/94/EU includes aspects related to objectives, planning, construction, and management of infrastructures for alternative fuels. It also identifies alternative fuels and sets mandatory objectives for electricity and natural gas and other discretionary objectives for hydrogen and LPG.

Despite decarbonization policies, CO_2 emissions from transport have increased since 1990 levels despite a decline between 2008 and 2013. In fact, in 2017, 27% of the total European Union (EU) Greenhouse gas emissions (GHG) came from the transport sector (including international aviation and maritime emissions).

Until now, most energy transition policies have focused on the electricity and industrial sectors, and consequently, there is still a great challenge in sectors such as buildings and transportation.

So far, the evolution of transportation was related to technological improvements in the sector. However, the energy transition processes have set objectives of reduction of greenhouse gases, the development of renewable energies and others that suppose relevant challenges to advance toward a low-carbon economy.

Given the relevance of the impact on greenhouse gas emissions and on other pollutant gases from transport and the likely increase of transport demand, transportation is on the focus of the energy transitions' policies.

This document addresses first the energy transitions' concept and then the implementation of related policies in some European countries to conclude on the imperative need to focus now on transportation. The next section examines European transport policies, especially making a review of the situation of alternative energies in the case of passengers' road transport.

The third section is dedicated to reviewing the main alternative energy sources for transportation, including economic and environmental aspects. The paper goes then on the situation and perspectives of the deployment of these energies in some European countries (the same whose energy transitions processes have been analyzed before) to finalize with a summary and some conclusions.

2 Energy Transitions. Concepts, European Approach, and Relevant Cases

Although there is no unique definition of the energy transition, the concept may be addressed by looking at the meanings of the two words.

By transition, the change from one state or situation to another may be understood. That is to say from one place to another place. This change also implies an understanding of going from one point to another being both known points. If the word energy is added to transition, the state or situation of energy in both points has to be considered when analyzing energy transitions.

For [69], "the term energy transition is used most often to describe the change in the composition (structure) of primary energy supply, the gradual shift from a specific pattern of energy provision to a new state or an energy system."

For the same author, "there are many energy transitions whose origins, progress, and accomplishments can be studied on levels ranging from local to global." In this respect, Smil refers to transitions to new energy sources, to a higher share of primary energies consumed in a secondary form as electricity, to the diffusion of new fuel and electricity energy converters.

The study of energy transitions in history reveals that many aspects of those energy transitions and the conclusions vary from the analysis at a global level to the level of the evolution and changes of energy mix or energy structures in different countries. Overall, it seems that the identification of two particular moments in history determines the points of origin and destination, depends on the approach of the historical analysis, as there are no predetermined dates and objectives of percentages of certain types of energies (i.e., coal, gas, and renewables).

These considerations may be relevant if we look at the European Union. It should be said that the point of origin of energy transitions in Europe might be fixed in 1996 when the green book on energy policy of the European Community was published and the basis for the determination of the objectives in renewables is set.

The objectives of the European Union for the years 2020 and 2030 are well known, in terms of GHG, the share of renewables in final energy consumption and improvement in energy efficiency. The quantification of objectives for the year 2030 and the relationship among the three objectives may be seen in the following Fig. 1.

The three objectives are very much related to the transport, as there are objectives for the penetration of renewables in energy use in the conventional vehicles (i.e., biofuels), and there is regulation related to the energy efficiency of Internal Combustion Engines (ICE) in terms of improving GHG specific emissions (gCO_2/km) and decreasing GHG emissions in transport.

Before going to the general policies for road transport in Europe, it is worthwhile to refer to the broader concept of mobility and asses briefly the global European context and trends.

For the EEA "*The mobility system spans all resources structures and activities involved in moving physical objects, including both people and goods. It is a complex system shaped by a multitude of forces, including economic and societal ones, such as cultural norms and lifestyles, evolving over long time scales*" [20].

"The transport sector is generally defined as an economic activity." "In contrast, the mobility system includes aspects that go beyond the economic activity, such as personal mobility and individual behaviour, infrastructures, urban and regional planning, investments, policy, and regulatory measures, as well as a multitude of actors such as producers, users, policymakers and civil society" [20].

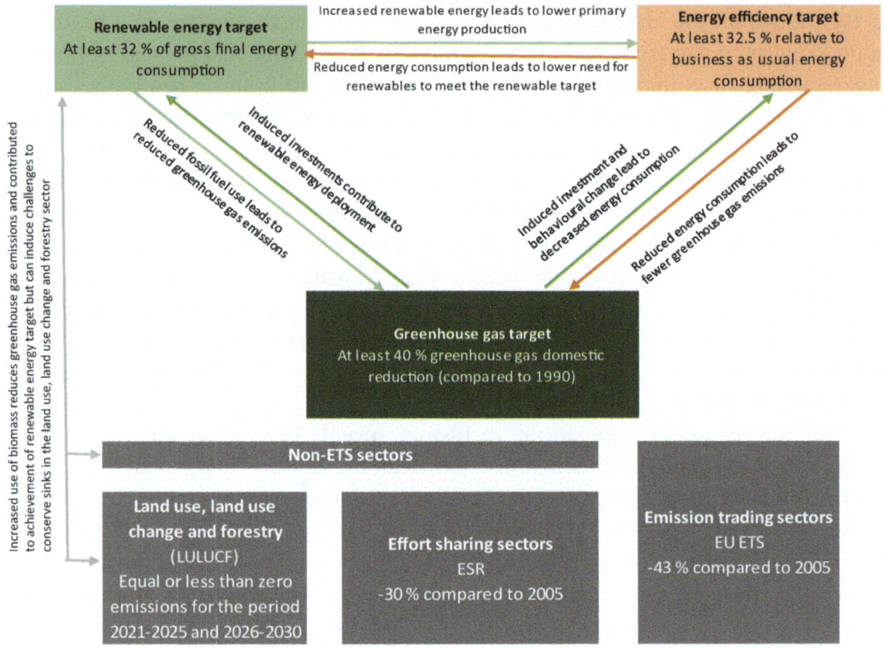

Fig. 1 Interaction between energy and climate objectives to 2030 *Source* EEA [22]

Transport accounted for 33% of the EU's final energy consumption in 2016 [18] and only 7% of the final energy used in transport came from renewable sources [20].

GHG emissions for transport accounted for 19.2% in 2016 and had increased by 26% since 1990. According to the European Commission, passenger and freight transport are expected to grow by about 42% and 60%, respectively, by 2050 compared with 2010 levels [17].

Therefore, it seems clear the relevance of transport in terms of GHG emissions and the importance to set transport within the concept of mobility. Furthermore, if fundamental changes in transport are necessary, those must be considered not only in the framework of mobility. It is also necessary to examine the drivers of change in the global European context.

The following figure illustrates six drivers of change. As may be seen later, all drivers have to do with transport; namely the growing urbanizing, climate change, global competition for resources (applicable to the materials of batteries for electric vehicles), accelerating technological change, power shifts in the global economy, and diversifying values and lifestyles (Fig. 2).

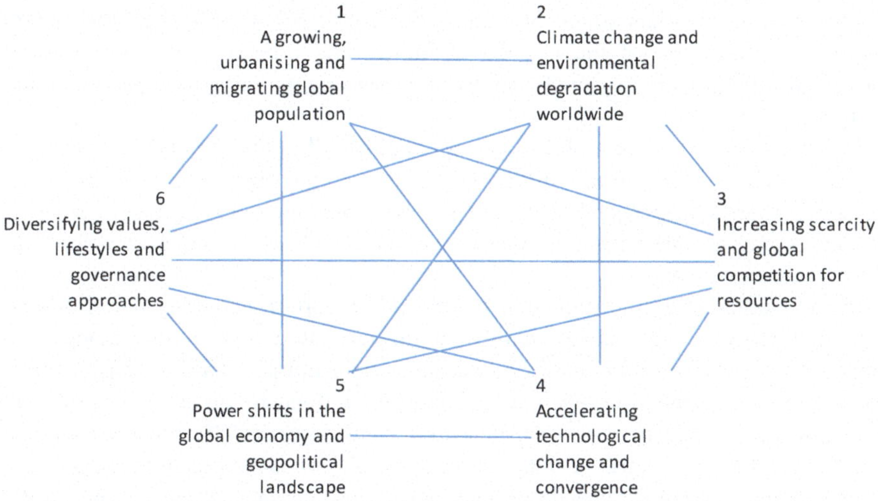

Fig. 2 Clusters of drivers of change *Source* Own elaboration from EEA [20]

2.1 Cases: Germany, France, Netherlands, Norway, The UK, and Spain

The concept of the energy transition is broad, and although energy transitions in Europe have more or less the same final objectives, the way to approach is different from country to country.

The energy transition in Germany, widely known as the "Energiewende," is the country's planned transition to a low-carbon economy, without relying on nuclear energy. Until recently, the German energy transition has focused on the electricity sector (mainly with the development of wind and solar) that poses significant challenges to the power system. Nevertheless, it is now extending the scope of its transition. It aims to power heating and transport with renewable energy, to replace fossil fuels entirely (which will have considerable implications for its carmakers, freight industry, or gas companies) and to reduce the energy consumption of the world's fourth-largest economy, by increasing efficiency both in households and in industry.

In parallel, the government considers that renewable energy-based hydrogen and other green gases are becoming an alternative to electricity [74]. In fact, in November 2019, the federal government announced the national strategy for hydrogen [66].

In France, the French energy transition for green growth Law of August 2015, marked the roadmap to mitigate climate change and diversify the energy mix, reducing the share of fossil fuels and nuclear power in favor of renewable energies. However, nuclear energy will remain at the heart of the country's short-/medium-term energy and environmental policy, to guarantee supply security [5].

This process also has among its pillars the use of fiscal tools (e.g., the carbon price), the promotion of the circular economy and efficiency in consumption; active participation by regional and local governments, safeguarding economic competitiveness.

In December 2019, the mobility law was published in the Official Journal. This law aims to transform the mobility policy with a single objective: make everyday transport easier, cheaper, and cleaner [61, 62]. It also includes the end of sales of fossil fuels cars by 2040, the deployment of electric recharging and the development of low-emission areas.

In the Netherlands, the transition toward a low-carbon economy is considered a gradual process that focuses mainly on the reduction of CO_2 emissions [48], supporting innovation and seizing economic opportunities. This transition introduces energy functionalities such as replacing natural gas by other sources of heat, redesigning industrial processes, carbon capture and storage (CCS), solar and wind in electricity, biofuels, electric vehicles, and strict emission standards in transport [72].

To achieve these objectives, the selection of measures include among others, the shift in energy transition, the closure of all coal-fired plants by 2030 at the latest, a national CO_2 price floor and the phasing out of Groningen gas field by the same year. In terms of transport, it is considered that there are limited opportunities to improve energy conservation in transport (sustainable driving habits, car sharing, and using lighter materials and engines that are more efficient). More far-reaching energy savings can be made by changing the types of vehicles and fuels (electric vehicles, biofuels, and biogases).

In the European arena, the Netherlands is committed to the implementation of stricter CO_2 emissions requirements for road transport and proposes stricter international requirements for shipping and aircraft emissions [63].

Since 2008, the main goal of energy-climate policies in Norway has been the reduction of GHG emissions. In 2016, the government presented a white paper on Norway's energy policy focusing on energy security, climate change, and industrial development. There are five priority areas for Norway's climate policy: reducing emissions from the transport sector, strengthening its role as a supplier of renewable energy, the development of low-emission industrial technology and clean production technology, environmentally sound shipping, and carbon capture, and storage [50]. In this context, a high level of public spending on energy RD&D and active efforts to develop carbon capture and storage are very welcome [55].

Transportation is one of the priorities in Norway, with a National Transport Plan (2014–2023) to incentivize public transport in urban areas. In the new National Transport Plan (2018–2019), the government is giving higher priority to improve the railways, to promote a shift from road to rail and low and zero-emission technology in the shipping industry [51]. At the same time, this plan incentivizes walking, cycling, and the use of transport in Norway's most significant towns; and it also establishes ambitious targets for phasing in zero-emission vehicles. For instance, all cars and buses sold in 2025 in the cities will be zero-emission vehicles by 2025 and most other vehicles will be zero-emission in 2030 [51].

In Spain, the energy transition from regulation has been in standby as a consequence of the political situation, but this will change in 2020 with the new government. In November 2018, this Ministry circulated a draft for a law on "Climate Change and Energy Transition." Sometime later, in February 2019, the Ministry for Ecological Transition published the "Strategic framework of Energy and Climate." This framework consists of three key documents for the energy transition. The first one is the draft of the law of "Climate Change and Energy Transition." The second is the draft of the "Integrated National Energy and Climate Plans 2011–2030" and the third one the "Strategy for a Fair Transition."

In the draft of the law, several issues are addressed, namely the objectives for 2030 and 2050; mobility without emissions (by 2040 vehicles should emit zero grams of CO_2 per kilometer); the no granting of licenses for exploration and production of hydrocarbons and the banning of hydraulic fracturing; measures for a fair energy transition, as well as measures related to the adaptation to climate change and finally the research, development, and innovation [25].

The government of the UK has transformed the energy policy during the last decades to achieve its goal to reduce GHG emissions by 100% by 2050 from 1990. Primarily it has continued its leadership on climate action, implementing the Electricity Market Reform, and strengthening policies on security of supply. In parallel, it has made progress on decarbonizing heat and electrifying transport [54]. The country has been able to transit toward a low-carbon economy in the electricity sector (replacing coal with gas and renewables) however; there is a need to make progress in terms of heating and transport.

The transport sector has only reduced its emissions by 3.2% from 1990 to 2018 [57]. Therefore, in 2017, the government published its 15-year renewable transport fuel strategy. One year later, in 2018 the government launched the Road to Zero Strategy to lead the world in zero-emission vehicle technology which "*sets out ambition for at least 50%—and as many as 70%—of new car sales to be ultra-low emission by 2030, alongside up to 40% of new vans*" [47]. This one is a technology-neutral strategy, as the government has no plan to ban any particular technology except diesel and gasoline by 2035.

Tables 1 and 2 include some of the main relevant targets in terms of energy transition process policies for the countries mentioned above, the first for 2030 and the second one for 2050.

There are energy transition processes with their singularities in other countries such as Sweden, who aspires to become one of the world's first fossil-free developed nations. The Swedish government set the objective of no net emissions of GHG into the atmosphere by 2045 and after that achieving negative emissions. Besides, it established a target to reduce emissions from domestic transport by at least 70% by 2030 compared to 2010. All this shall need strong local and regional climate efforts [25].

As can be observed, transportation does not seem to be a priority or to be on the focus of energy transitions mentioned, being, however, one of the sectors that contribute significantly to GHG emissions.

Table 1 Main objectives of the energy transitions by 2030

	Germany	France	Netherlands	Norway	UK	Spain
GHG emissions (compared to 1990)	Minimum −55%	−41%	−49%	−40%	−1.765Mt (2028–2032)	−21%
RES in gross final energy consumption	30%	34%	32% (bandwidth of 27–35%)	–	–	42%
RES in gross electricity consumption	Minimum 50%	40%	–	–	–	74%
RES in the transport sector	–	15%	–	–	12.4% (in 2032)	–
Reduction in primary energy consumption (compared to 2008)	−20% (in 2020, no objective for 2030)	−24.6% (a reference to 2007)	32.5%	–	–	–
Reduction in final energy consumption (compared to 2007)	–	20%	–	–	–	39.6%
Reduction in final energy consumption in transport	−10% (in 2020, no objective for 2030)	–	–	–	–	–
Emissions reduction in transport	–	–	−7.3%	–	–	–

Source Own elaboration from Appunn and Wettengel [6], Ministère de la transition écologique et solidaire [61, 62], Álvaro Hermana and Larrea Basterra [5], Government of the Netherlands [49], Magnus commodities [59], Government.no [50] and IEA [54]

Table 2 Main objectives of the energy transitions by 2050

	Germany	France	Netherlands	Norway	UK	Spain
GHG emissions (compared to 1990)	Largely GHG neutral −80–95%	−75%	−95%	Low-emission society	−100%	−90%
RES in gross final energy consumption	60%	–	–	–	–	–
RES in gross power consumption	Minimum 80%	–	–	–	–	100%
RES in the transport sector	–	–	–	–	–	–
Reduction in primary energy consumption (compared to 2008)	−50%	–	–	–	–	–
Reduction in final energy consumption (compared to 2012)	–	−50%	–	–	–	–
Reduction in final energy consumption in transport	−40%	–	–	–	–	–
Emissions reduction in transport	–	–	−60%	–	–	–

Source Own elaboration from Appunn and Wettengel [6], Álvaro Hermana and Larrea Basterra [5], Government of the Netherlands [49], Government.no [50], Euro-CASE [25], Larrea Basterra and Bilbao Ozamiz [57] and Ministère de la transition écologique et solidaire [61, 62]

3 Road Transport in Europe—General Policies

At the European level, normative and legislation related to transport is very wide and cover quite a variety of issues. As has been seen in the previous chapter, transport can be considered and is a part of the mobility concept. Furthermore, transport includes several modes such as rail, air, sea, and road. However, this chapter on policies and regulations focuses only on road transport and especially on passengers' transportation.

3.1 Strategies and Policies Related to Mobility

Policies, strategies, and regulations related to transport have become along the time more holistic and more inclusive. In this respect, it is very significant the communication of 2017 "Europe on the move. An agenda for a socially fair transition towards clean, competitive and connected mobility for all" [32].

This document specifies that "*The Energy Union Strategy of February 2015 identified the transition to an energy-efficient, decarbonized transport sector as one of its key areas of action, and the 'Clean Energy for all Europeans' package of November 2016 included action to accelerate the deployment of low-carbon transport fuels and to support electro-mobility. The measures which were already outlined in the 'Strategy for Low-Emission Mobility' adopted in July 2016 are now being implemented.*"

In the European strategy for low-emission mobility, there are three main issues related to the regulatory framework: firstly, "*optimizing the transport system and improving its efficiency*"; secondly, "*scaling up the use of low-emission alternative energy for transport*"; and thirdly, "*moving towards zero-emission vehicles.*"

In the last two points, three issues are identified, namely the effective framework for low-emission alternative energies, the roll-out of infrastructure for alternative fuels and the interoperability and the standardization for electromobility. Concerning zero-emission vehicles, the Commission has started to work on revising the post-2020/2021 carbon dioxide standards for cars, vans, lorries, buses, and coaches [32].

The points as mentioned earlier should be put in the context of the European's ambition that "*must be to make rapid progress towards having a clean, competitive and connected mobility system integrating all means of transport in place by 2025.*"

Once road transport has been put in the context of mobility regulation, we should address regulation related to road transport. In this respect, three issues shall be referred, namely (a) limits of pollutant emissions following among others, euro 5 and euro 6 regulations, (b) renewable energies in transport, and (c) alternatives energies in transport.

(a) *Pollutant emissions limits*

Since 1991, there are clear regulations at the EU level related to pollutants, such as CO, NO_x, PM and ultrafine particulates, as may be seen in Table 3.

In this respect, it is essential to mention the Regulation 715/2007/EU on type-approval of motor vehicles concerning emissions from light passenger and commercial vehicles (Euro 5 and Euro 6) and on access to vehicle repair and maintenance information [37].

Another relevant document is Regulation (EC) 595/2009 of the European Parliament and of the Council of 18 June 2009 on type-approval of motor vehicles and engines concerning emissions from heavy-duty vehicles (Euro VI) and on access to vehicle repair and maintenance information [41]. It was later modified by the Commission regulation 2011/582/CE that determines emission limits for heavy-duty vehicles [35].

Table 3 Summary of the main pollutant sectors

	EU		Spain	
Pollutant	Sector	A percentage of total emissions (%)	Sector	Percentage from total emissions (%)
SO_x (Sulfur oxides)	Electricity generation and distribution	59	Electricity generation	50
NO_x (Nitrogen oxides)	Road transport	39	Road transport	32
PM_{10} (Particulate matter 10)	Residential, commercial, and institutional	42	Residential, commercial and institutional	32
$PM_{2,5}$ (Particulate matter 2,5)	Residential, commercial, and institutional	57	Residential, commercial and institutional	42
NMVOC (Non-methane Volatile organic compound)	Industrial processes and products' use	50	Solvents	48
CO (carbon monoxide)	Residential, commercial, and institutional	47	Residential, commercial and institutional and waste	Both 28
NH_3 (Ammonia)	Agriculture	94	Agriculture	96

Note Percentages of anthropic emissions (natural ones are excluded). For Spain, the agriculture and livestock emissions are presented together
Source Modified and translated from Álvarez Pelegry et al. [3]

Directive 2009/33/EC on the promotion of clean and energy-efficient road transport vehicles "*requires contracting authorities, contracting entities as well as certain operations to take into account lifetime energy and environmental impacts, including energy consumption and emissions of* CO_2 *and certain pollutants, when purchasing road transport vehicles with the objectives of promoting and stimulating the market for clean and energy-efficient vehicle and improving the contribution of the transport sector to the environment, climate and energy policies of the Community*" [39].

In the same year was adopted the Regulation (EC) 443/2009 setting emission performance standards for new passenger cars as part of the Community's integrated approach to reduce CO_2 emissions for light-duty vehicles [40].

This regulation imposes obligations on each manufacturer of passenger cars to ensure that average specific emissions targets are not exceeded. Since 2012, there has been an increasing percentage of passenger cars to which limits applicable have evolved from 65% in 2012 to 100% from 2015 onward. Limits vary with the mass of the car, and there are both super-credits for new passenger cars with less than 50 g CO_2/km emissions; and an "excess emissions premium" (EEP). The EEP has different values depending on the time (from 2012 to 2018 and from 2019). From 2019 excess emissions have to pay 95 €/gCO_2/km, being the excess emissions calculated following a formula that takes into account 130 gCO_2/km and a coefficient that applies to the difference in mass over a mass reference.

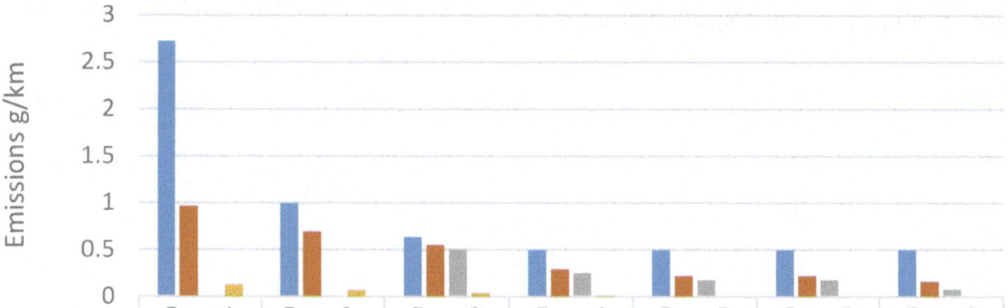

	Euro 1 (1992)	Euro 2 (1996)	Euro 3 (2000)	Euro 4 (2005)	Euro 5 (2009)	Euro 5 (2011)b	Euro 6 (2014)
CO	2.72	1	0.64	0.5	0.5	0.5	0.5
HC+NOx	0.97	0.7	0.56	0.3	0.23	0.23	0.17
NOx	0	0	0.5	0.25	0.18	0.18	0.08
PM	0.14	0.08	0.05	0.025	0.005	0.005	0.005

CO HC+NOx NOx PM

Graph 1 Limits established for diesel engines *Source* Álvarez Pelegry and Menéndez Sánchez [2]

Taking into account the formula, each manufacturer has different limits. The estimations for European manufactures in most of them show that they should have to pay premium credits of US$39,000 million[1] given the gap between the limits and the actual performance of emissions. This gap may have effects on dropping the sales of vehicles with higher specific emissions and selling more electric vehicles so to obtain emissions more in line with the limits.

A few years later 2009, the European Commission published the Commission Regulation (EU) 2016/427 of 10 March 2016 amending Regulation (EC) No 692/2008 as regards emissions from light passenger and commercial vehicles (Euro 6). This last regulation established the procedures for testing vehicles under real conditions in the road [29].

Because of the European regulation, the evolution in terms of limits for the CO, HC, NO_x, and PM may be seen in Graphs 1 and 2.

(b) *Renewables in transport*

There are four important pieces of legislation to be referred chronologically related to renewables in transport. The first one is the Directive 2009/28/EC of the European Parliament and of the Council of 23 April 2009 on the promotion of the use of energy from renewable sources, known as the RED Directive. The second one is the Directive 2009/30/EC of the European Parliament and of the Council of 23 April 2009 amending Directive 98/70/EC as regards the specification of petrol, diesel, and gas-oil

[1] Dawson and Sachgau [13] and Muñoz and Galetovic [64].

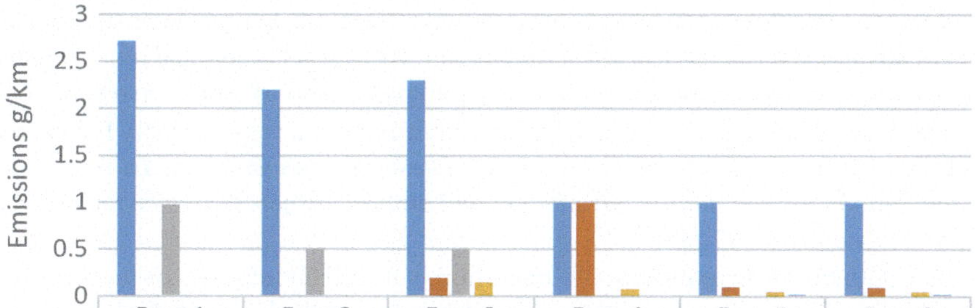

	Euro 1 (1992)	Euro 2 (1996)	Euro 3 (2000)	Euro 4 (2005)	Euro 5 (2009)	Euro 6 (2014)
CO	2.72	2.2	2.3	1	1	1
HC	0	0	0.2	1	0.1	0.1
HC+NOx	0.97	0.5	0.5	0	0	0
NOx	0	0	0.15	0.08	0.06	0.06
PM	0	0	0	0	0.005	0.005

CO HC HC+NOx NOx PM

Graph 2 Limits set for gasoline engines *Source* Álvarez Pelegry and Menéndez Sánchez [2]

and introducing a mechanism to monitor and reduce greenhouse gas emissions and amending Council Directive 1999/32/EC as regards the specification of fuel used by inland waterway vessels and repealing Directive 93/12/EEC[2] (called the Fuel Quality Directive, FQD) [38]. The third document is the Directive (EU) 2015/1513 of the European Parliament and of the Council of 9 September 2015 amending Directive 98/70/EC relating to the quality of petrol and diesel fuels and amending Directive 2009/28/EC on the promotion of the use of energy from renewable sources[3] (so-called targets and named "ILUC Directive").

Moreover, the fourth one is Directive (EU) 2018/2001 of the European Parliament and of the Council of 11 December 2018 on the promotion of the use of energy from renewable sources [44].

The RED Directive "*obliges the Member States to achieve a general target of 20% renewables in all energy used by 2020 and a sub-target of 10% renewables in the transport sector. The EU Member States are required to meet a minimum binding target of 10% renewable energy share in the transport sector by 2020. All types of renewable energy used in all transport modes are included in the target setting*" [58].

[2] European Parliament and of the Council [38].

[3] European Parliament and of the Council [43].

Some renewable energy sources are counted differently. For instance, the contribution of advanced biofuels[4] toward achieving the 10% target is counted twice, whereas electricity from renewable energy sources for road transport counts five times [58].

The Fuel Quality Directive (FQD) sets environmental requirements for gasoline and diesel to reduce their GHG intensity. "*These requirements consist of technical specifications for fuel quality parameters and binding targets to reduce the fuels' life cycle GHG emissions*" [58].

By 2020, the FQD requires a 6% reduction in the GHG intensity of fuels traded in the EU. The FQD places the responsibility for reducing the life cycle GHG emissions of fuels traded in the EU on fuel suppliers.

The European Directive, 2015/1513, completes and revises the 2009 RED and FQD Directives. Among the main elements of the ILUC Directive, it: (a) tackles indirect land-use change emissions through a 7% cap on conventional biofuels, including biofuels produced from energy crops, to count toward the renewable energy directive targets regarding final consumption of energy transport in 2020. Member states can set a lower cap. The Directive also (b) sets an indicative 0.5% target for advanced biofuels as a reference for national targets which will be set by EU member states in 2017 and (c) harmonizes the list of feedstocks for biofuels across the EU, whose contribution would count double toward the 2020 target of 10% for renewable energy in transport. Next, it (d) requires that biofuels produced in new installations emit at least 60% fewer GHG than fossil fuels and (e) introduces stronger incentives for the use of renewable electricity in transport (by counting it five more times for renewable electricity in road transport and 2.5 times in rail). Finally the Directive (f) includes several additional reporting obligations for the fuel providers, EU countries, and the European Commission.

As mentioned, more recently, the Directive UE 2018/2001 (so-called RED II) has been enacted. Its article 25 states that "*To mainstream the use of renewable energy in the transport sector, each Member State shall set an obligation on fuel suppliers to ensure that the share of renewable energy within the final consumption of energy in the transport sector is at least 14% by 2030 (minimum share).*"

This RED II Directive also states that "*The Commission shall assess that obligation (*at least 32% of renewables collectively in the Union's gross final consumption of energy by 2030), *intending to submit, by 2023, a legislative proposal to increase it in the event of further substantial costs reductions in the production of renewable energy, where necessary to meet the Union's international commitments for decarbonization, or were justified on the grounds of a significant decrease in energy consumption in the Union.*"

Within the minimum share referred above, the contribution of advanced biofuels and biogas produced from feedstock[5] as a share of final consumption of energy in the transport sector shall be at least 0.2% in 2022, at least 1% in 2025 and at least 3.5% in 2030. Article 26 sets limits to the share of biofuels, bioliquids as well as biomass

---

[4] According to the RED, biofuels must meet minimum sustainability criteria as well as minimum GHG savings per energy unit.

[5] Listed in Part A of Annex IX of the Directive.

fuels consumed in transport produced from food and feed crops, with a maximum of 7%.

(c) *Alternative energy vehicles*

In 2013, the European Commission released the communication "Clean Power for Transport: A European alternative fuels strategy." The document recognizes that "*While further efficiency improvements spurred by EU regulations on vehicle emissions of* CO_2 *will continue to represent the lowest hanging fruits in the short term to medium term, low* CO_2 *alternatives to oil are also indispensable for a gradual decarbonization of transport, a key objective of the Europe 2020 strategy for smart, sustainable and inclusive growth, towards a target of a 60% reduction of* CO_2 *emissions for transport by 2050*" (set out in the 2011 White Paper on Transport) [27].

The Commission recognizes that "*Initiatives to support alternative transport fuels exist at both EU and national level but a coherent and stable overarching strategy with an investment-friendly regulatory framework needs to be put in place.*" It also notes that "*previous European initiatives supporting alternative fuels, including market quota and favourable taxation, have been followed up in uneven and disjointed ways.*"

The communication identified a comprehensive mix of alternative fuels: LPG, natural gas including biomethane, LNG, CNG, and GTL[6] (Gas to Liquids), electricity, biofuels, liquid, and hydrogen and establishes that the strategy should not give preference to any particular fuel, thereby keeping technology neutrality.

In the priorities for further action, the Commission identified the following (a) addressing alternative fuels infrastructure, (b) developing standard technical specifications, (c) addressing consumer acceptance, and (d) addressing technological development.

In October 2014 the Directive 2014/94/EU of the European Parliament and of the Council of 22 October 2014 on the deployment of alternative fuels infrastructure was published, also referred as DAFI Directive [42].

This Directive established a common framework of measures for the deployment of alternative fuel infrastructure in the European Union to minimize dependence on oil and to mitigate the environmental impact of transport. It sets out "*minimum requirements for the building-up of alternative fuels infrastructure, including recharging points for electric vehicles and refuelling points for natural gas (LNG and CNG) and hydrogen to be implemented through Member States' national policy frameworks, as well as common technical specifications for such recharging and refuelling points and user information requirements.*"

Alternative energies considered in this Directive are electricity, hydrogen, biofuels,[7] synthetic and paraffinic fuels, natural gas including biomethane (in gaseous and liquefied forms), and liquefied petroleum products.

---

[6] Liquefied Natural Gas, Compressed Natural Gas and Gas to Liquids.

[7] As defined in point (i) of Article 2 of Directive 2009/28/EC.

The relevance of the Directive is that it requires each Member State to adopt a national policy framework for the development of the market alternative fuels in the transport sector and the deployment of relevant infrastructure, that shall contain at least the following elements: (a) an assessment of the current state and future development of the market for alternative fuels in the transport sector, (b) national targets and objectives, (c) necessary measures to ensure that the national targets are reached, (d) measures that can promote the deployment of alternative fuels in public transport services, (e) designation of the urban/suburban agglomerations which subject to the market needs are to be equipped with recharging points or CNG refueling points, (f) an assessment of the need to install refueling points for LNG in ports, and (g) consideration of the need to install electricity supply at airports.

It is interesting to note that concerning hydrogen the Directive refers to "Member States which decide to include hydrogen refuelling points"; therefore the type of obligation about this alternative fuel seems different from others, such as electricity, biofuels, or gas.

In 2017, a communication from the Commission assessed the situation and the needs and suggested an action plan.[8] Some of the conclusions referred to the need to accelerate deployment in two areas, namely to implement the backbone infrastructure for the core network by 2025 the latest, and the need to ramp up infrastructure in urban and suburban areas, where vehicles are used for most of the time.

It also concluded that addressing the broader transportation network requires more considerable efforts and pointed out that the level of ambition between different Member States varied significantly. It also noted the enormous investments needed in infrastructures. Taking together the total estimated investment needs for publicly accessible alternative fuels infrastructure in the EU, the total amount reaches 5,200M€ by 2020 and additional 16,000M€ by 2025.[9]

By the end of 2018, the deployment of infrastructure for alternative fuels was considered insufficient in a report of the Committee on transport and tourism that called for action.[10]

More recently, in 2019, the Commission assessed the Member States National Policy Frameworks[11] (NPFs) and included a methodology for the assessment mentioned and carried out an overview of targets, objectives, and level of attainments from all NPFs and examined the overall contribution of NPFs to EU policy targets.[12]

The main conclusion is that there are considerable differences in the various NPFs of the different Member States. Furthermore, some states have not sent the

[8] European Commission [31].

[9] In particular, electricity up to 904 million euros (M€) by 2020, CNG up to 357M€ by 2020, LNG up to 275M€ by 2025 for LNG road vehicles, hydrogen up to 707M€. The enumeration here referred to is not complete.

[10] Ertug [24].

[11] NPFs should include national targets for the deployment of alternative fuels infrastructure in the respective Member State.

[12] European Commission [34].

plans to the European Commission yet. There are also considerable levels of implementation, including the lack of completion of some plans. All in all, the degree of implementation is not satisfactory. Fig. 3 shows a summary of the above mentioned regulation.

4 Alternative and Conventional Energies. Economic and Environmental Aspects

As has been previously noted, "*Whilst further efficiency improvements spurred by EU regulations on vehicle emissions of* CO_2 *will continue to represent the lowest hanging fruits in short to medium term, low-*CO_2 *alternatives to oil are also indispensable for gradual decarbonization of transport*" [28].

As has been mentioned, there is still a long way to go in terms of decarbonization of transport, and each country's point of depart is different. In fact, "*there is no single fuel solution for the future of mobility and all alternative fuel options must be pursued, with a focus on the needs of each transport mode.*"

As a consequence, a strategic approach to meet the long-term needs of all transport modes must be built on a comprehensive mix of alternative fuels [28].

Some of the main energy alternatives to conventional fuels (such as diesel and gasoline) in road transport are Liquefied Petroleum Gas, natural gas, electricity, biofuels, hydrogen, and E-fuels. Each energy alternative has its advantages and disadvantages in economic and environmental terms. However, there are also relevant technological questions such as batteries for electric vehicles.

LPG is a by-product of the hydrocarbon fuel chain, composed mainly by propane and butane. Vehicles that use this type of energy have a high degree of autonomy, around 480 km [45].

Natural gas can be supplied from fossil fuels, from biomass and waste as biomethane and in the future, from methanization of hydrogen.[13] Besides natural gas may be as compressed natural gas or liquefied.[14] Vehicles propelled by natural gas may be mono fuel or biofuel. In this second case, vehicles have two deposits (one for natural gas and other for the conventional fuel). There are also dual-fuel vehicles. The autonomy of natural gas vehicles is less than conventional vehicles but more significant than the electrical ones (300–800 km) [12].

The technology of electric vehicles is maturing quickly. One key component of this kind of vehicle is the battery both in economic and environmental terms.[15]

"*Biofuels can be produced from a wide range of feedstock through technologies in constant evolution and used directly or blended with conventional fossil fuels.*

[13] Natural gas can be transformed to a liquid fuel by first decomposing it to a "synthesis gas," consisting of hydrogen and carbon monoxide, and then by refining to a synthetic fuel, fully compatible with existing combustion engines and fuel infrastructure [26].

[14] Liquefied Natural Gas is more used in heavy vehicles and ships [14, 68].

[15] For more information, see chapter 4: Batteries for electric vehicles.

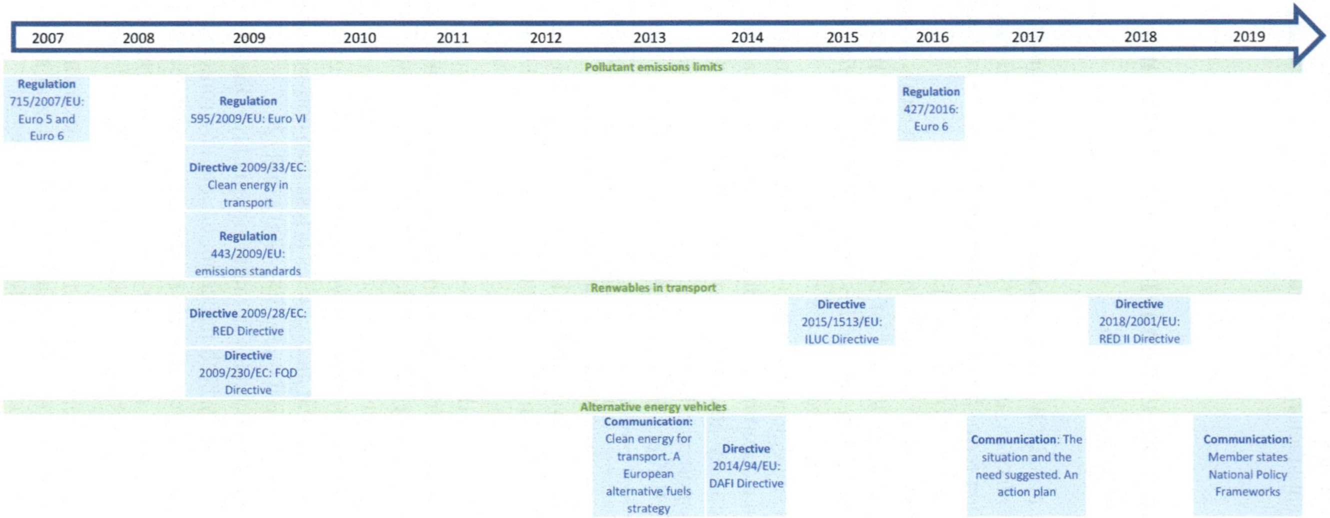

Fig. 3 Some relevant directives, regulations and communications related to road transport in Europe (not exhaustive) *Source* Own elaboration

They include bioethanol, biomethanol, and higher bio alcohols, biodiesel (fatty acid methyl ester, FAME), pure vegetable oils, hydrotreated vegetable oils, dimethyl ether (DME), and organic compounds." [28]. Their great advantage is that they can be mixed with conventional fuels and that the regulation in Europe has promoted them to reduce oil product dependence.

Hydrogen is a universal energy carrier and can be produced from primary energy sources. The technology related to "*hydrogen fuel cell vehicles is maturing, and is being demonstrated in passenger cars, city buses, light vans, and inland ship applications.*" [28]. The Alternative Fuels Infrastructure Directive mentions specifically hydrogen and establishes that the Member States that opt for hydrogen must deploy infrastructure by end-2025.

E-fuels are synthetic fuels resulting from the combination of "green or e-hydrogen" produced by electrolysis of water with renewable electricity and CO_2 *captured either from a concentrated source (e.g. flue gases from an industrial site) or from the air (via direct air capture, DAC). E-fuels are also described in the literature of electrofuels, power-to-X (PtX), power-to-liquids (PtL), power-to-gas (PtG) and synthetic fuels* [10].

After carrying out some basic considerations concerning alternative fuels, there is a need to compare some economic and environmental key parameters.

4.1 Economic Aspects

Among the key economic aspects for the penetration of alternative transportation fuels are the price of the fuels, the costs of the refueling points, and the price of the vehicles themselves.

The price of alternative fuels is a quite complicated issue, as there are substantial differences in their prices among countries. The differences across countries are due in part to the various taxes and subsidies. Therefore, all countries have access to the same oil and natural gas prices from international markets but then have decided to impose different taxes [46].

For instance, the situation for LPG, as a general rule, is that richer countries have higher prices while poorer countries and the countries that produce and export natural gas have significantly lower prices. The case of electricity is different as it depends on regional markets and energy mixes, among others. Consequently, it must be analyzed for each case.

In the case of hydrogen, e.g., prices ranged in 2015 from US$12.85 to more than US$16 per kilogram (kg), but usually, it is US$13.99 per kg (equivalent on a price per energy basis to US$5.60 per gallon of gasoline), which translates to an operating cost of US$0.21 per mile [16] [8].

---

[16] While future price is uncertain, NREL estimates that hydrogen fuel prices may fall to the $10–$8 per kg range in the 2020–2025 period [9].

In terms of refueling points, some new infrastructure investments would be needed in the case of LPG. These investments will be higher for the recharging points for electric vehicles and CNG [4].

The cost of electricity charging points on public roads was in 2017 in the range of €7,500–€10,000 for conventional charging and €35,000–€50,000 for fast charging. For home charging points, with power levels between 3.7 and 22 kW, the cost was between €2,200 and €2,400 per point. For CNG refueling stations, costs varied the same year depending on the capacity and filling type (slow or fast): from a minimum of US$5,000 to a maximum of US$700,000 [4].

For biofuels, there is no need for additional investment in infrastructures as there is already a supply infrastructure in place. The situation is different for hydrogen and e-fuels as refueling stations will need electricity connections for the pre-cooling facilities [67]. In this case, the price level of the refueling infrastructure is similar to the infrastructure cost for electric vehicles or even higher in the short term (by 2020); however, it is expected that this cost will decrease faster than the cost of electric recharging stations [52].

As far as the price of vehicles is concerned, for conventional ones (gasoline and diesel) prices are around €14,000–16,000 per vehicle [4].

Taking into consideration market prices, EVs were priced from €27,000 to €95,000 in 2019 [53] and CNG vehicles at €26,000 in 2014 [11].

The price of compressed natural gas vehicles has decreased since 2014 and are in the range of CNG vehicles as Table 4 shows [56]. Conventional hybrid costs were between €22,000 and 115,000 in 2019 [53]. LPG vehicle prices are in the range from 15,000 to 47,000 [7]. The case of hydrogen vehicles is different as there are no so

Table 4 A comparison of the prices of different alternative fuels vehicles (€)

	Electricity	LPG	CNG	PHEV	H_2
Minimum price	27,200	15,050	14,190	38,100	68,000
Model	h, four-seater	Arona 1.0 TGI 90 S&S Reference Edition 6 V	Mii Ecofuel	225 xe	Nexo
Manufacturer's brand	Mini	Seat	Seat	BMW	Hyundai
Maximum price	95,500	46,840	46,830	163,855	66,000
Model	Model X	A5 Sportback 40 g-tron 170 Aut. 7 V	Audi A5 Sportback g-tron	Autobiography	Mirai (hydrogen fuel cell vehicle)
Manufacturer's brand	Tesla	Audi	Audi	Land Rover	Toyota
Price range	68,300	31,790	32,640	125,755	2,000

Source Own elaboration

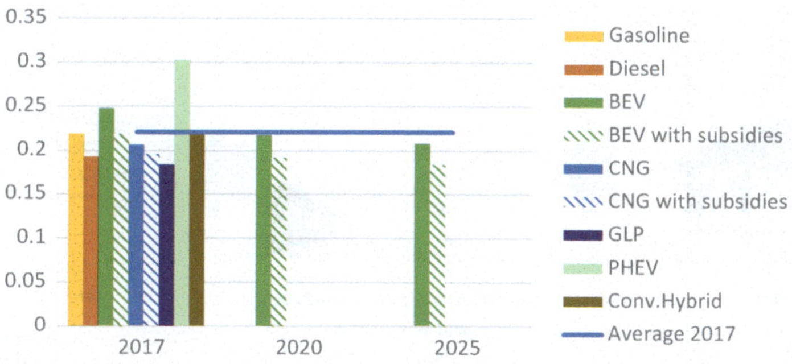

Graph 3 Comparative evolution of estimated TCO (€/km) *Source* Álvarez Pelegry et al. [4]

many vehicles in the market today. Some of them are not available in all countries [73].

In economic terms, comparing alternative energies vehicles need to take into consideration the "Total Cost of Ownership for the owner" (TCO) that includes not only vehicles' prices but the cost of fuel, insurance, and maintenance over its lifetime as well.

Graph 3 shows the TCO of different types of vehicles. It may be seen that by 2025 it could be expected that TCO of alternative fuels vehicles shall be similar to those of conventional ones (gasoline and diesel).

More recent estimates of electric vehicles' TCO also coincide in stating that by mid of this decade total price of ownership cost be similar among the electric vehicles and the conventional ones and that by the end of this decade will be the turn of hydrogen and hydrogen fuel cells' TCO [52, 65].

Nowadays, along with conventional vehicles, natural gas and LPG ones are sufficiently proven technologies with high production volumes. However, EV technology (in particular batteries[17]) remains on the learning curve. Therefore, future reductions in battery prices may affect the TCO, and it is expected a reduction of more than 50% over the coming decade of the price of batteries[18] [60].

4.2 *Environmental Aspects*

Comparing environmental issues of alternative fuels and fuels for conventional vehicles, the first distinction is between air pollutants and greenhouse gas emissions as they operate on a different scale of impact and potential damage. *Air pollutant emissions have a more significant direct impact when people are exposed to them at the local level, and their main risk is related to health when they are inhaled.*

[17] For more information, see chapter 4: Batteries for electric vehicles.

[18] Battery's cost represented in 2014 35% of the price of electric vehicles.

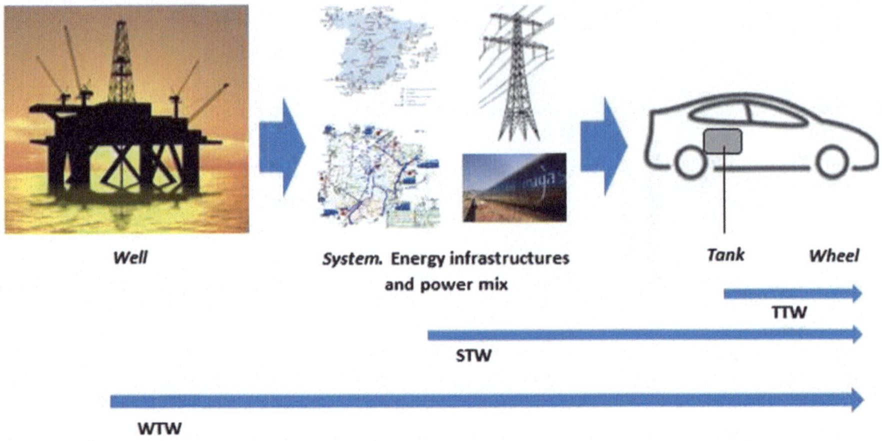

Fig. 4 Comparison of TTW, STW, and WTW *Source* Álvarez Pelegry and Menéndez Sánchez [2]

On the other hand, the GHG emissions present the global risk of climate change but do not represent a direct or immediate problem for citizens [4].

Each category of transportation emissions should then be analyzed within the frame of different scales (or emissions cycle), depending on their origin and the geographical reach of their potential damage. A smaller scale (or shorter cycle)—used in the case of air pollutants—the relevant parameters measure pollutant emissions from the tank to the wheel (TTW) that represents only those emissions that are generated on vehicle roads.[19] *A more global scale—as in the case of GHGs—covers the entire chain of emissions. Known as from well to wheel (WTW), this scale includes not only the emissions directly from the vehicle in situ but also the production, treatment, and transportation of the fuel before it reaches the vehicle* [4]. Both TTW and WTW emissions scales are necessary critical to understand the broader environmental implications of each fuel.

Furthermore, at the country level, the emissions depend on the structure and nature of the national energy systems and are in between WTW and WTT. This consideration is especially relevant for the analysis of the electric vehicle, given that its environmental impact (i.e., emissions reductions) is directly related to the structure of the national power mix and the level of emissions resulting from electricity generation. This parameter is called from *system to wheels* (STW). This is illustrated in Fig. 4.

Both CO_2 and air pollutant emissions in the TTW, STW, and WTW calculations vary by type of energy. As both the STW and WTW measures for battery electric vehicles (and also partly for hybrid vehicles, when it is recharged) depend on the emissions of the particular national electricity generation mix, such estimates of emissions levels will probably change in the coming years, given the trend to decarbonize

[19] Not only the emissions produced from the combustion of fuel should be considered, but also those produced by the erosion of the wheels and the road when the vehicle is moving (which throws particulate pollution into the atmosphere).

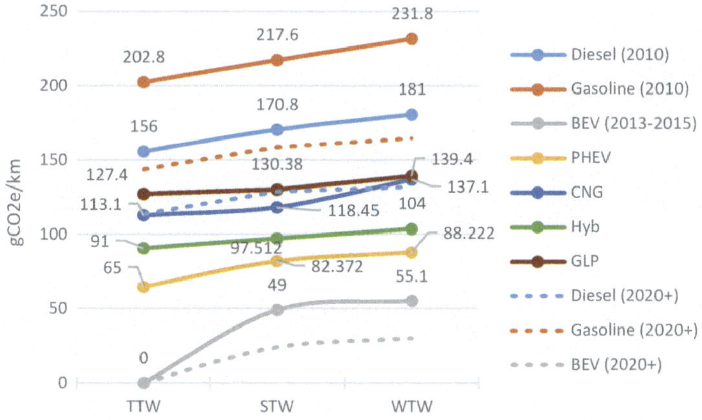

Graph 4 CO_2e emissions (TTW, STW, and WTW) for each vehicle type *Note* Diesel, gasoline, and AFV 2020 + emissions are: Diesel (gCO_2e/km): TTW 114, STW 129, WTW 132. Gasoline (gCO_2e/km): TTW 144, STW 159, WTW 165. BEV (gCO_2e/km): STW 24, WTW 30. *Source* Álvarez Pelegry and Menéndez Sánchez [2]

the power sector. On the other hand, the implementation of the Euro 6 regulation shall imply emissions reductions for diesel and gasoline vehicles. Graph 4 illustrates CO_2 emissions for different fuels and technologies for the parameters TTW, STW, and WTW.

As far as pollutant emissions of NO_x and particulate matters, the following graph shows the comparison of different alternative energies and for conventional (gasoline and diesel vehicles). As may be seen the 2020+ projections for air pollutant emissions foresee reductions for battery electric vehicles, but plug-in hybrids electric vehicles would also result in net pollutant emissions reductions (to the extent that they rely on charging) (Graph 5).

It must be noticed that the metrics of the parameters WTW, STW, and TTW do not cover the full lifecycle. Being the life cycle assessment (LCA) of growing relevance, given the implications of sustainability in the energy sector, the next section deals with this topic for the case of electric vehicles, considering electric batteries.

4.3 *Batteries for Electric Vehicles*

Having seen the comparisons in terms of WTW, STW, and TTW for different pollutants and CO_2 emissions, it is relevant to examine one of the critical components of the electric vehicle, namely the battery. Different studies deal with this topic and focus on the environmental implications of batteries from their LCA and make comparisons with conventional vehicles.

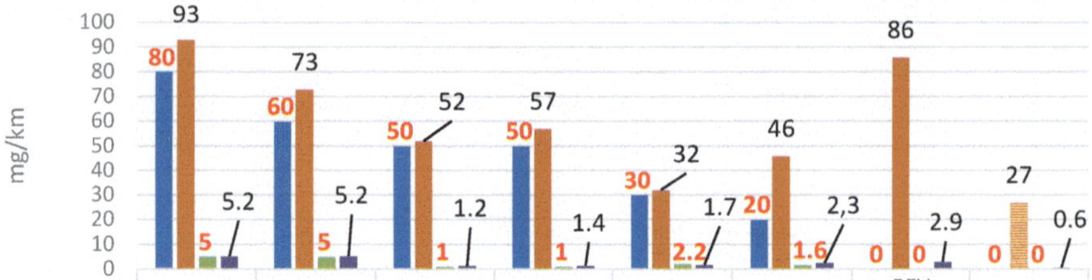

	Diesel	Gasoline	LPG	CGN	Hyb	PHEV	BEV (2013-2015)	BEV (2020+)
NOx TTW	80	60	50	50	30	20	0	0
NOx STW	93	73	52	57	32	46	86	27
PM TTW	5	5	1	1	2.2	1.6	0	0
PM STW	5.2	5.2	1.2	1.4	1.7	2.4	2.9	0.6

NOx TTW NOx STW PM TTW PM STW

Graph 5 Pollutant emissions, TTW, and STW, by vehicle type *Note* 1 The particulate emissions which derive from electricity consumption are known as PM_{10}. *Note* 2 Substantial reductions of NO_x are foreseen in the electric system for the years to come. *Source* Álvarez Pelegry et al. [4]

Thomas [70] analyzes battery electric vehicles (BEV) with lithium-ion traction batteries looking to the emissions along its life cycle that includes the mineral supply, battery production, rest of BEVs production, use, and recycling (Graph 6).

The results of the study mentioned above show that BEV has higher production emissions than the internal combustion engine vehicles (ICEV) that are attributed to the traction battery. However, BEV moderately reduces GHG emissions compared to both diesel and gasoline vehicles assuming a 150,000 km vehicle lifetime.

It is important to note that "*The expected decarbonization of the European power sector, coupled with improvements in electrochemical and powertrain performances,*

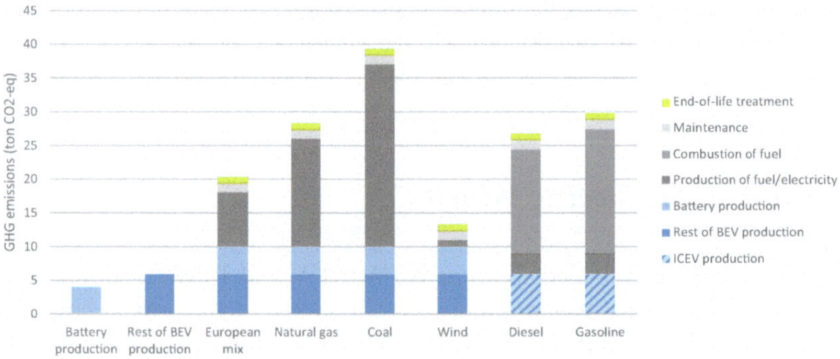

Graph 6 Life cycle GHG emissions of mid-sized 24 kWh battery electric (left) and internal combustion engine (right) vehicles *Note* The vehicle's operational lifetime is assumed to be 150,000 km. *Source* Replicated from Thomas [70]

should lead to an increase in the environmental benefits of BEV use over time, particularly from a climate change perspective" [70].

The same author also considers that as the batteries grow in size, BEV may increase the driving range and consumer acceptance, nevertheless, their production will require more resources and as a consequence will lead to higher emissions. Therefore, there is a need to strike the right balance between battery and vehicle size.

Another recent report of the European Environment Agency carried out an in-depth consideration of the environmental impact of batteries using LCA as well as taking a broader "circular economy" approach. The environmental impacts are grouped in climate change, health, and ecosystem [19].

Here below we shall deal briefly with the first two documents.

In relation with climate change impacts, the study states that "*across its life cycle, a typical BEV in Europe offers a reduction in GHG emissions compared with its equivalent ICEV.*" The size of the vehicles, the electricity mix, and whether the BEV is compared with a petrol or diesel vehicle are some of the factors that can introduce differences in the results.

Human health impacts include air pollution, noise exposure, and "human toxicity." BEV can offer local air quality benefits due to zero exhaust emissions; however, they still emit particulate matter (PM). As far as noise is concerned, the difference in noise emissions between BEVs and ICEVs depends on vehicle speeds. Noise in urban areas where speeds are generally low have some benefits, but it is unlikely to be a substantial benefit on rural roads or motorways. Concerning human toxicity impacts, the literature on climate change impacts is limited; however, it suggests that BEV impacts could be higher overall than their ICEV equivalents.

The relevance of the battery value chain for sustainable development and climate change mitigation is demanding growing attention. The vision to 2030 of the WEF and GBA incorporates three elements (a) a circular battery value chain as a major driver to meet the Paris Agreement, (b) the transformation of the economy creating new jobs and additional economic value, and (c) a value chain safeguarding human rights, supporting a just energy transition and fostering economic development, in line with the UN Sustainable Development Goals. Still, as the study points out "*it will, however, not be achieved without a dynamic shift from the current development trajectory. This change requires immediate actions by companies, investors and policy-makers, in consultation with all stakeholders.*" [75].

5 Situation and Perspectives of the Penetration of Alternative Energies

Alternative energies are essential to reduce the impact of transport on the environment. Nevertheless, to achieve this goal, there is a need to deploy them and develop measures to stimulate the shift from conventional vehicles (diesel and gasoline) to alternative fuels vehicles.

In general terms, future projections of alternative energies vehicles present a progressive growth, with notable or substantial increases in sales by 2030. World sales were expected to be between two and five million vehicles in 2020 [1].[20]

According to the same source, for 2030, in general, the electric vehicle fleet will be multiplied by ten or more compared to 2020. The total park figures in the year 2030 could be between 80 and 228 million units. That is, there is significant variability in the estimates.

Once putting to perspective some global figures for alternative energies and in particular to electric vehicles, this chapter shall deal mainly with the present development of alternative energies in some European countries.

5.1 *Europe: The EU*

To encourage the expansion of alternative fuels vehicles, as has been seen, the DAFI Directive requires that Member States provide a minimum infrastructure for alternative fuels such as electricity, LPG, natural gas, or hydrogen. In the case of electricity, Member States must develop public recharging points by 2020, at least in urban and suburban agglomerations. The DAFI Directive considers the need of one recharging point per ten electric vehicles; however, this is not a binding requirement.

For compressed natural gas vehicles, the Directive requires that the Member States must ensure a sufficient number of publicly accessible refueling points with common standards and recommends a minimum of one refueling point every 150 km by the end of 2025, not being this a binding objective.

The Directive also aims to ensure a sufficient number of publicly accessible refueling points with common standards for hydrogen as for compressed natural gas, which should be built by the end of 2025 again.

The Directive mentioned above set that Member States have to notify to the European Commission their National Policy Frameworks (NPF) including targets, objectives, and measures for the development of alternative energies, and the development of infrastructure.

Taking into account the requirements of the DAFI Directive and each country's point of depart, some key data that reflect the situation for EU-28 in 2019, can be observed in the following Table 5.

In 2019, 4.02% of the total fleet of passenger cars were from alternative energies, most of them (3.06%) were LPG vehicles. LPG fueled most of the alternative energies vehicles (76%), 12% by natural gas, 6% were battery electric vehicles, and 6% PHEV.

The number of LPG vehicles in 2019 was 7.9 million in Europe. It is the most relevant energy alternative, and it is distributed quite homogenously in different European countries. The relative facility to install refueling points at the existing

[20] See Table 3, page 28.

Table 5 Current situation of alternative fuels' vehicles and related publicly accessible infrastructure in the European Union (2019)

Basic data	Population	50,80,00,000			
	Passenger cars	26,00,60,286			
	Highway (km)	1,31,718			
Infrastructure (number of recharging or refueling points publicly accessible)	Electricity	1,84,609			
	H_2	138			
	LPG	34,239			
	Natural gas (CNG)	3,722			
Alternative energies (number of vehicles)	Electric vehicles 2020	6,47,778	0.25%	6%	
	CNG vehicles 2020	12,12,161	0.47%	12%	
	H_2 vehicles 2025	922	0.00%	0%	
	Other fuels (LPG)	79,67,583	3.06%	76%	
	PHEV	6,26,032	0.24%	6%	
	Total	1,04,54,476	4.02%		232.7048979

Source Own elaboration from EAFO [16]

conventional petrol stations facilitates the deployment of this energy. There were 232.7 vehicles per refueling station.

One of the most critical points for the deployment of compressed natural gas is the need for refueling stations. Although the grid for transmission and distribution of natural gas is quite developed in Europe, it is not the case with the refueling stations, as the actual numbers in some of the European countries show that the deployment is lagging behind objectives, save in some countries where can be considered that the degree of deployment is enough (i.e., Germany and France[21]).

In 2019, there were 647,778 electric vehicles in the European roads and 626,032 PHEV. The number of models is increasing, and the original equipment manufacturers have announced that by 2021 the number of electric vehicle models in the market will triple [71].

It is considered that the lack of recharging points with a universal plug is a significant obstacle to market uptake. The actual numbers, as provided in the report of the European Commission, indicate that in 2019 there were twenty-six fast public charging points (>22 kW) per 100 km highway and seven electric vehicles per charging point [15].

[21] European Commission [30].

5.2 *Cases: Germany, France, Netherlands, Norway, The UK, and Spain*

The deployment of alternative energies vehicles has been heterogeneous in different Member States. Table 6 collects key data and main objectives, set by each country related to alternative fuels' vehicles and infrastructure, for the countries of the second chapter of this document, which are: Germany, France, Netherlands, Norway, UK, and Spain.

Table 7 shows the situation of deployment of both alternative energies vehicles and their related infrastructure by the end of the year 2019. Then there are some comments on the evolution of the accomplishment of the objectives.

Norway and the Netherlands are the countries where the penetration of alternative fuels is higher (11.5 and 3.7%, respectively). In Spain, this percentage is the lowest among the countries analyzed (only 0.7%).

The highest number of battery electric vehicles are in Norway, France, and Germany. However, the market share of this type of alternative fuel is more important in Norway (7.8%) and the Netherlands (0.92%) while it is around 0.3 and 0.5% in Germany and France.

In absolute terms, LPG vehicles are the most relevant ones (with 39% of total alternative fuels vehicles), even if in Norway there are no LPG vehicles since 2017. In fact, since 2008 the number of LPG vehicles in Norway has decreased.

LPG and battery electric vehicles are the alternative fuels vehicles with a higher level of penetration. Compressed natural gas has only achieved a market share higher than 0.1% in Germany and the Netherlands.

As has been seen, Spain has the least number of alternative fuels vehicles among the countries considered, however, the objectives could be on their way to be achieved, especially in CNG (40.7%) and in LPG vehicles (51.1%). The Netherlands, in 2019, had already more than half of the battery electric vehicles that must be on the roads by 2030 (55.3%). Germany and France are farther from achieving the objective of electric vehicles (14.8 and 16% respectively). The United Kingdom has already an electricity fleet of 21% of the final objective for 2030.

However, the countries referred before have already accomplished their objectives in terms of ad hoc infrastructure for electric, CNG, and H_2 vehicles.[22] The infrastructure is less developed in Spain, where only LPG and H_2 vehicles have already expanded their network of refueling stations comparing to the objectives. This situation means that perhaps the countries are advancing on energy alternatives infrastructure objectives, even if the degree of development would be not enough to transform the present fleet of vehicles into another more sustainable. In the Netherlands and the United Kingdom, the electric and CNG infrastructure is less evolved than in the remaining Member States.

Behind this deployment, different measures could be classified in different groups: tax benefits, local benefits, subsidies, financial incentives; such as tax exemptions

[22] In the case of France LPG infrastructure has already been developed, and H_2 infrastructure is on its way to being already deployed (83% of the 2030 goal).

Table 6 A future estimate of alternative fuels' vehicles and related publicly accessible infrastructure set in the NPF of different member States

	Germany	France	Netherlands	Norway	UK	Spain
Population	83.019.213	67.028.048	17.282.163	5.328.212	66.647.112	46.934.632
Passenger cars	47.095.784	33.020.132	8.373.244	2.700.000	34.887.915	24.074.216
Highways (km)	13.009	11.618	3.055	523	3.803	15.523
Number of vehicles						
Electric vehicles 2020	1,000,000	960,000	140,000	–	396,000–431,000	38,000–150,000
CNG vehicles 2020	–	–	–	–	–	17,200
H_2 vehicles 2025	–	–	2,120 (by 2020)	–	–	500 (by 2020)
Other fuels	–	–	–	–	–	200,000–250,000 (LPG by 2020)
Publicly accessible infrastructure						
Electric vehicles 2020	43,000	35,000	17,844	–	12,000–13,500	–
CNG vehicles 2020	913	79/210	145	–	8 (NPF)/13 (EAFO)	76
H_2 vehicles 2025	400 (maximum)	11 (NPF)/9 (EAFO)	20 (by 2020)	–	65	20
Other fuels	–	1,750 (LPG vehicles)	–	–	–	800 (LPG vehicles)

Note 1 EAFO = European alternative fuels observatory
Note 2 For information on estimates of the future fleet of electric vehicles see [1]
Source Own elaboration from Álvarez Pelegry [33]

Table 7 Current situation of alternative fuels' vehicles and related publicly accessible infrastructure in different Member States (2019)

	Germany	France	Netherlands	Norway	UK	Spain
Number of vehicles						
Electric vehicles 2020	148.086	153.695	77.392	211.796	86.777	24.180
CNG vehicles 2020	86.013	2.382	10.244	222	23.000	7.000
H_2 vehicles 2025	207	128	173	180	157	1
Other fuels (LPG)	383.409	138.000	132.536	0	120.000	115.000
PHEV	117.893	54.481	96.046	98.374	160.715	17.280
Total	735.608	348.686	316.391	310.572	390.649	163.461
Publicly accessible infrastructure (recharging and filling stations)						
Electric vehicles 2020	32.704	29.538	50.289	12.473	27.204	8.622
CNG vehicles 2020	868	137	211	25	18	1.100
H_2 vehicles 2025	75	12	3	5	14	5
Other fuels (LPG)	7.361	1.600	1.389	92	1.164	120
Total	41.008	31.287	51.892	12.595	28.400	9.847

Source Own elaboration from EAFO [16]

from the annual circulation tax for a while, tax exemption in the company tax, registration tax benefits, and VAT benefits. Other measures include free parking and reserved parking slots, toll exemption on regional highways for electric vehicles, traffic lanes reserved. There are also purchase benefits in some countries.

In the case of the infrastructure among the measures can be found especially economic incentives and incentives for infrastructure purchasing.

6 Summary and Conclusions

The transition toward a low-carbon economy is a reality in a quite number of countries and particularly in Europe. Energy transitions in Europe during the last decade have relied on objectives in three main areas: decrease of GHG emissions, increase of the share of renewables, and improvement of energy efficiency, being interrelated.

Unlike other energy transitions that have taken place in the past, energy transitions nowadays are driven by international agreements and regulation. Most of the advancements and developments have focused on the electricity sector. Consequently, much remains to be done in other sectors, such as the transportation sector. The cases of Germany, France, Netherlands, Norway, Spain, and the UK confirm these statements.

Mobility systems are complex, shaped by a multitude of forces. All drivers of change, as identified by the European Environmental Agency, have to do with transport. All accounts to say that transport is a complex system that has to be addressed holistically, being alternative fuels, a pivotal element to contribute to better transport systems.

The European Union has a considerable number of levers to impulse changes in transport systems. From the regulation of the quality of fuels to limits in emissions of GHG and pollutants and the rather recent Directive on alternative energies infrastructures (DAFI Directive). These policies have been enacted in a variety of Directives and Regulations. Furthermore, the European Commission made public strategies and communications which approach transport, not only from mobility and transport systems but also concerning innovation and industry.

As mentioned before the EU has developed an important legislative acquis, which highlights, among others the DAFI Directive. This Directive considers the relevance of the ad hoc infrastructure and sets a series of measures to reach the objectives of alternative energies vehicles for the next years. Nevertheless, other measures can be taken to deploy these technologies.

In this regard, alternative fuels are a necessity to meet the objectives by 2050 in both the reduction of GHG emissions and the perspectives of transport demand growth. Alternative fuels include, inter alia: biofuels, synthetic and paraffinic fuels, electricity, natural gas, (including biomethane) in gaseous form (compressed natural gas) and liquefied form (liquefied natural gas), liquefied petroleum gas, and hydrogen.

In terms of economics and environmental issues, these fuels have advantages and disadvantages (the technological aspects are not considered in this paper). In economic terms, some concepts should be considered, such as the price of the vehicles, the cost of the infrastructure and the total cost of ownership.

In environmental terms, there are differences among vehicles if GHG emissions and pollutant emissions are considered. Other differences appear when taking into account from what stage of the alternative fuel the environmental impact is calculated (such as WTW, TTW, and STW). Moving to a zero-emission circular economy, with a focus on well-to-wheel rather than tailpipe emissions, will have a growing interest, and LCA, considering the full life cycle.

Each European country has adopted different objectives and measures. At present, the most relevant alternative energy among the EU countries is LPG. However, the number of electricity and PHEV vehicles has increased over time, and even if the most relevant developments are in Norway and the Netherlands, there are positive perspectives in countries as Germany and France. In these countries, the passenger cars' fleet is more significant, and there is a real need to advance toward low-carbon transport.

In some other European countries that have been analyzed, the infrastructure needed for alternative fuels is being developed relatively fast taking into consideration the objectives, as most of the objectives set for the future have already been accomplished.

In any case, there is a real need to develop more refueling and recharging infrastructure, considering the principle of technology-neutral strategy. This need is particularly the case for those alternative energies that at present have a low level of development such as electricity, natural gas, and hydrogen (in those countries, which have opted for these alternatives).

References

1. Álvarez Pelegry E (2019) Transición energética y transporte. El papel de las energías alternativas y las convencionales. Econ Ind (411):21–34. https://www.mincotur.gob.es/Publicaciones/Publicacionesperiodicas/EconomiaIndustrial/RevistaEconomiaIndustrial/411/%C3%81LVAREZ%20PELEGRY.pdf
2. Álvarez Pelegry E, Menéndez Sánchez J (2017) Energías alternativas en el transporte de pasajeros por carretera. El caso de la CAPV: análisis y recomendaciones. Bilbao. www.orkestra.deusto.es
3. Álvarez Pelegry E, Menéndez Sánchez J, Bravo López M (2018) Calidad del aire. Situación en España y escenarios a futuro para la CAPV, Madrid y Barcelona. Bilbao. Orkestra. https://www.orkestra.deusto.es/images/investigacion/publicaciones/informes/cuadernos-orkestra/calidad-del-aire.pdf
4. Álvarez Pelegry E, Menéndez Sánchez J, Larrea Basterra M (2017) Sustainable mobility in the European Union: alternative fuels for passenger transport. In: Isbell P, Álvarez Pelegry E (eds), pp 55–90. Washington, DC. https://archive.transatlanticrelations.org/wp-content/uploads/2017/11/Energy-and-Transportation-in-the-Atlantic-Basin-Text-FINAL-final.pdf
5. Álvaro Hermana R, Larrea Basterra M (2018) La transición energética en Francia. Bilbao. https://www.orkestra.deusto.es/es/investigacion/publicaciones/informes/cuadernos-orkestra/1642-transicion-energetica-francia
6. Appunn K, Wettengel J (2020) Germany's greenhouse gas emissions and climate targets. https://www.cleanenergywire.org/factsheets/germanys-greenhouse-gas-emissions-and-climate-targets
7. Autofácil (2020) Coches GLP. https://cochesnuevos.autofacil.es/coches-glp.html
8. California Energy Commission (2015) Joint agency staff report on assembly bill 8: assessment of time and cost needed to attain 100 hydrogen refueling stations in California. California. https://ww2.energy.ca.gov/2015publications/CEC-600-2015-016/CEC-600-2015-016.pdf
9. California Fuel Cell Partnership (2015) Cost to refill. https://cafcp.org/content/cost-refill
10. Concawe (2019) A look into the role of e-fuels in the transport system in Europe (2030–2050). In: Literature review, vol 28, 1st edn. https://www.concawe.eu/wp-content/uploads/E-fuels-article.pdf
11. CR (2014) The natural-gas alternative. The pros & cons of buying a CNG-powered car. https://www.consumerreports.org/cro/2012/03/the-natural-gas-alternative/index.htm
12. CREG (2018) Study on the cost-effectiveness of natural gas (CNG or compressed natural gas) used as fuel in cars. Belgium. https://www.creg.be/sites/default/files/assets/Publications/Studies/F1736EN.pdf
13. Dawson C, Sachgau O (2019) Europe's tough emissions rules come with $39 billion threat. https://www.bloomberg.com/news/articles/2019-06-26/europe-s-tough-new-emissions-rules-come-with-39-billion-threat
14. DNV-GL (2020) LNG as ship fuel. https://www.dnvgl.com/maritime/lng/index.html
15. EAFO (2019) Country detail. https://www.eafo.eu/countries/european-union/23640/summary
16. EAFO (2020) Vehicles and fleet. https://www.eafo.eu/countries/european-union/23640/vehicles-and-fleet
17. EC (2017) Communication from the Commission to the European Parliament, the European Council, the Council, the European Economic and Social Committee, the Committee of the Regions and the European Investment Bank investing in a smart, innovative and sustainable industry. A renewed EU industrial policy strategy (COM), vol 479. https://eur-lex.europa.eu/legal-content/EN/TXT/?uri=COM:2017:0479:FIN
18. EC (2018) UN climate talks: EU plays instrumental role in making the Paris agreement operational. https://ec.europa.eu/clima/news/un-climate-talks-eu-plays-instrumental-role-making-paris-agreement-operational_en
19. EEA (2018) Electric vehicles from life cycle and circular economy perspectives. TERM 2018: transport and environment reporting mechanism (TERM) report, Luxembourg. https://www.eea.europa.eu/publications/electric-vehicles-from-life-cycle/electric-vehicles-from-life-cycle/viewfile#pdfjs.action=download

20. EEA (2019a) The European environment—State and outlook 2020 knowledge for transition to a sustainable Europe, Luxembourg. https://doi.org/10.2800/96749, https://www.eea.europa.eu/publications/soer-2020
21. EEA (2019b) Greenhouse gas emissions from transport in Europe. https://www.eea.europa.eu/data-and-maps/indicators/transport-emissions-of-greenhouse-gases/transport-emissions-of-greenhouse-gases-12
22. EEA (2019c) Trends and projection in Europe 2019-tracking progress towards Europe climate and energy targets
23. EMSA (2020) Alternative fuels. http://www.emsa.europa.eu/main/air-pollution/alternative-fuels.html
24. Ertug I (2018) Report 1.10.2018. On deployment of infrastructure for alternative fuels in the European Union: time to act! (2018/2023(INI)), Brussels. http://www.europarl.europa.eu/doceo/document/A-8-2018-0297_EN.html
25. Euro-CASE (2019) Energy transitions in Europe. Common goals but different paths. https://energiesysteme-zukunft.de/en/publications/euro-case-report/
26. European Commission (2013a) Commission staff working document. Actions towards a comprehensive EU framework on LNG for shipping, Brussels. https://eur-lex.europa.eu/LexUriServ/LexUriServ.do?uri=SWD:2013:0004:FIN:EN:PDF
27. European Commission (2013b) Communication from the Commission to the European Parliament, the Council, the European Economic and Social Committee and the Committee of the Regions clean power for transport: A European alternative fuels strategy. COM/2013/017 final, Brussels. https://eur-lex.europa.eu/legal-content/EN/ALL/?uri=CELEX%3A52013PC0017
28. European Commission (2013c) Communication from the Commission to the European Parliament, the Council, the European Economic and Social Committee and the Committee of the Regions. Clean power for transport: a European alternative fuels strategy. {SWD(2013) 4 final}, Brussels. https://eur-lex.europa.eu/legal-content/EN/TXT/PDF/?uri=CELEX:52013PC0017&from=GA
29. European Commission (2016) Commission regulation (EU) 2016/427 of 10 March 2016 amending regulation (EC) no 692/2008 as regards emissions from light passenger and commercial vehicles (Euro 6), Brussels. https://eur-lex.europa.eu/legal-content/EN/TXT/?uri=CELEX%3A32016R0427
30. European Commission (2017a) Summary on national plans for alternative fuel infrastructure. https://ec.europa.eu/transport/sites/transport/files/2017-11-08-mobility-package-two/summary_of_national_policy_frameworks_on_alternative_fuels.pdf
31. European Commission (2017b) Commission staff working document. Detailed assessment of the national policy frameworks, accompanying the document Communication from the Commission to the European Parliament, the Council, the European Economic and Social Committee and the Committee of the Regions "towards the broadest use of alternative fuels-an action plan on alternative fuels infrastructure under article 10(6) of Directive 2014/94/EU, including the assessment of national policy frameworks under article 10(2) of Directive 2014/94/EU", Brussels. https://eur-lex.europa.eu/resource.html?uri=cellar:d80ea8e8-c559-11e7-9b01-01aa75ed71a1.0001.02/DOC_1&format=PDF
32. European Commission (2017c) Communication from the Commission to the European Parliament, the Council, the European Economic and Social Committee and the Committee of the Regions. Europe on the move. An agenda for a socially fair transition towards clean, competitive and connected mobility for all, Brussels. https://eur-lex.europa.eu/legal-content/EN/TXT/PDF/?uri=CELEX:52017DC0283&from=EN
33. European Commission (2019a) Alternative fuels. Member state Country Fiches. https://ec.europa.eu/transport/sites/transport/files/alternative-fuels-ms-country-fiches.pdf
34. European Commission (2019b) Commission staff working document. Report on the assessment of the Member states national policy frameworks for the development of the market as regards alternative fuels in the transport sector and the deployment of the relevant infrastructure pursuant to article 10 (2) of Directive 2014/94/EU, Brussels

35. European Commission (2011) Commission regulation (EU) no 582/2011 of 25 May 2011 implementing and amending regulation (EC) no 595/2009 of the European Parliament and of the Council with respect to emissions from heavy-duty vehicles (euro VI) and amending Annexes I and III to Directive 2007/46/EC of the European Parliament and of the Council, Brussels. https://eur-lex.europa.eu/legal-content/EN/TXT/PDF/?uri=CELEX:32011R0582&from=ES
36. European Partliament and European Council (2009a) Directive 2009/29/EC of the European Parliament and of the Council of 23 April 2009 amending Directive 2003/87/EC so as to improve and extend the greenhouse gas emission allowance trading scheme of the Community (2009). http://eur-lex.europa.eu/LexUriServ/LexUriServ.do?uri=CELEX:32009L0029:EN:NOT
37. European Parliament and European Council (2007) Regulation (EC) no 715/2007 of the European Parliament and of the Council of 20 June 2007 on type-approval of motor vehicles with respect to emissions from light passenger and commercial vehicles (Euro 5 and Euro 6) and on access to vehicle repair and maintenance information, Brussels. https://eur-lex.europa.eu/legal-content/EN/ALL/?uri=CELEX%3A32007R0715
38. European Parliament and European Council (2009b) Directive 2009/30/EC of the European Parliament and of the Council of 23 April 2009 amending Directive 98/70/EC as regards the specification of petrol, diesel and gas-oil and introducing a mechanism to monitor and reduce greenhouse gas emissions and amending Council Directive 1999/32/EC as regards the specification of fuel used by inland waterway vessels and repealing Directive 93/12/EEC, Brussels. https://eur-lex.europa.eu/legal-content/EN/TXT/?uri=CELEX%3A32009L0030
39. European Parliament and European Council (2009c) Directive 2009/33/EC of the European Parliament and of the council of 23 April 2009 on the promotion of clean and energy-efficient road transport vehicles, Brussels. https://eur-lex.europa.eu/legal-content/EN/ALL/?uri=CELEX%3A32009L0033
40. European Parliament and European Council (2009d) Regulation (EC) no 443/2009 of the European Parliament and of the Council of 23 April 2009 setting emission performance standards for new passenger cars as part of the community's integrated approach to reduce CO_2 emissions from light-duty vehicles (text with EEA relevance), Brussels. https://eur-lex.europa.eu/legal-content/EN/TXT/?uri=CELEX%3A32009R0443
41. European Parliament and European Council (2009e) Regulation (EC) no 595/2009 of the European Parliament and of the Council of 18 June 2009 on type-approval of motor vehicles and engines with respect to emissions from heavy-duty vehicles (euro VI) and on access to vehicle repair and maintenance information and amending regulation (EC) no 715/2007 and Directive 2007/46/EC and repealing Directives 80/1269/EEC, 2005/55/EC and 2005/78/EC, Brussels. https://eur-lex.europa.eu/legal-content/EN/ALL/?uri=celex%3A32009R0595
42. European Parliament and European Council (2014) Directive 2014/94/EU of the European Parliament and of the Council of 22 October 2014 on the deployment of alternative fuels infrastructure, Brussels. https://eur-lex.europa.eu/legal-content/EN/TXT/PDF/?uri=OJ:L:2014:307:FULL&from=EN
43. European Parliament and European Council (2015) Directive (EU) 2015/1513 of the European Parliament and of the Council of 9 September 2015 amending Directive 98/70/EC relating to the quality of petrol and diesel fuels and amending Directive 2009/28/EC on the promotion of the use of energy from renewable sources, Brussels. https://eur-lex.europa.eu/legal-content/EN/TXT/?uri=CELEX%3A32015L1513
44. European Parliament and European Council (2018) Directive (EU) 2018/2001 of the European Parliament and of the Council of 11 December 2018 on the promotion of the use of energy from renewable sources, Brussels. https://eur-lex.europa.eu/legal-content/EN/TXT/?uri=uriserv:OJ.L_.2018.328.01.0082.01.ENG
45. Gasdrive (n.d.) All about LPG gas. http://www.gasdrive.co.za/all-about-lpg-gas/
46. Global Petrol Prices.com (2020) LPG prices, liter. https://www.globalpetrolprices.com/lpg_prices/
47. GOV.UK (2018) Government launches road to zero strategy to lead the world in zero-emission vehicle technology. https://www.gov.uk/government/news/government-launches-road-to-zero-strategy-to-lead-the-world-in-zero-emission-vehicle-technology

48. Government of the Netherlands (2019a) Climate agreement. The Hague
49. Government of the Netherlands (2019b) Climate policy. https://www.government.nl/topics/climate-change/climate-policy
50. Government.no (2019) Meld. st. 41 (2016–2017) Norway's climate strategy for 2030: a transformational approach within a European cooperation framework—meld. st. 41 (2016–2017) report to the Storting (white paper). https://www.regjeringen.no/en/dokumenter/meld.-st.-41-20162017/id2557401/sec2?q=bedrifter
51. Guldbrandsen Froysa K (2018) Energy transition in Norway and at UiB. University of Bergen. https://www.uib.no/sites/w3.uib.no/files/attachments/2018-06-04_renewable_energy_froysa_final.pdf
52. Hall D, Lutsey N (2019) White paper. Estimating the infrastructure needs and costs for the launch of zero-emission trucks. Washington, DC. https://theicct.org/sites/default/files/publications/ICCT_EV_HDVs_Infrastructure_20190809.pdf
53. Hayes J (2019) Your essential guide to buying an electric car in 2020. https://www.rte.ie/brainstorm/2019/1211/1098528-your-essential-guide-to-buying-an-electric-car-in-2020/
54. IEA (2019a) Energy policies of IEA countries, United Kingdom 2019 review. https://webstore.iea.org/energy-policies-of-iea-countries-united-kingdom-2019-review
55. IEA (2019b) Norway. https://www.iea.org/countries/norway
56. Jiménez J (2020) Todos los coches que funcionan con gas GLP y GNC en España. https://www.auto10.com/reportajes/todos-los-coches-que-funcionan-con-gas-glp-y-gnc-en-espana/16306
57. Larrea Basterra M, Bilbao Ozamiz M (2020) La transición energética en Reino Unido. Unpublished manuscript
58. Lonza L, Deix S, Maas H, Caiado C, Hamje HDC, Reid A (2016) JRC technical reports. EU renewable energy targets in 2020: 2015 legislative update JRC. https://www.concawe.eu/wp-content/uploads/2017/03/jrc103306_103306_20161220_jec_biofuels_study_pubsy.pdf
59. Magnus commodities (2017–2020) The Netherlands in 2030: (almost) halfway the elimination of greenhouse gasses. https://www.magnuscmd.com/the-netherlands-in-2030-almost-halfway-the-elimination-of-greenhouse-gasses/
60. McKinsey & Company (2014) Evolution. Electric vehicles in Europe: gearing up for a new phase? http://www.mckinsey.com/~/media/McKinsey%20Offices/Netherlands/Latest%20thinking/PDFs/Electric-Vehicle-Report-EN_AS%20FINAL.ashx
61. Ministère de la Transition écologique et solidaire (2019a) La loi mobilité. https://www.ecologique-solidaire.gouv.fr/loi-mobilites
62. Ministère de la transition écologique et solidaire (2019b) Projet de Plan national integre énergie-climat de la France. https://ec.europa.eu/energy/sites/ener/files/documents/france_draftnecp.pdf
63. Ministry of Economic Affairs of the Netherlands (2016) Energy report transition to sustainable energy, Netherlands. www.government.nl/ministries/ez
64. Muñoz CM, Galetovic A (2014) La Wende de la Energiewende (Alemania reconsidera su transición a las energías renovables). http://brevesdeenergia.com
65. Sauras A (2019) Cuánto cuesta producir y utilizar el hidrógeno para los coches. https://www.autofacil.es/coches-electricos-e-hibridos/2019/10/16/cuesta-producir-utilizar-hidrogeno-coches/52811.html
66. Schahinian D (2019) German federal government announces hydrogen strategy. https://www.hannovermesse.de/en/news/news-articles/german-federal-government-announces-hydrogen-strategy
67. Siegemund S (2017) "E-fuels" study. The potential of electricity-based fuels for low-emission transport in the EU. An expertise by LBST and dena, Berlin. www.dena.de
68. Smajla I, Karasalihović Sedlar D, Drljača B, Jukić L (2019) Fuel switch to LNG in heavy truck traffic. Energies. https://doi.org/10.3390/en12030515
69. Smil V (2010) Energy transitions: history, requirements, prospects. Praeger, Santa Barbara, California
70. Thomas M (2018) Research for TRAN committee-battery-powered electric vehicles: market development and lifecycle emissions. European Union. Brussels. Retrieved from https://www.europarl.europa.eu/RegData/etudes/STUD/2018/617457/IPOL_STU(2018)617457_EN.pdf

71. Transport and Environment (2019) Electric car models to triple in Europe by 2021—Market data. https://www.transportenvironment.org/press/electric-car-models-triple-europe-2021-%E2%80%93-market-data
72. Van't Hof W (2018) Energy transition in the Netherlands. phasing out of gas. Electricity Directorate: Ministry of Economic Affairs and Climate Policy. https://ec.europa.eu/energy/sites/ener/files/documents/01.b.02_mf31_presentation_nl-fuel_switch-vanthof.pdf
73. Viñuela S (2019) Todos los coches de hidrógeno que se venden en España. https://www.autobild.es/listas/todos-coches-hidrogeno-venden-espana-422951
74. Wettengerl J (2019) German government puts focus on green hydrogen for energy transition—Media. https://www.cleanenergywire.org/news/federal-government-puts-focus-green-hydrogen-energy-transition-media
75. World Economic Forum and Global Battery Alliance (2019) A vision for a sustainable battery value chain in 2030. Unlocking the full potential to power sustainable development and climate change mitigation. http://www3.weforum.org/docs/WEF_A_Vision_for_a_Sustainable_Battery_Value_Chain_in_2030_Report.pdf

The manufacturer's authorised representative in the EU is Springer Nature Customer Service Centre GmbH, Europaplatz 3, 69115 Heidelberg, Germany. If you have any concerns regarding our products, please contact ProductSafety@springernature.com

Printed and bound by CPI Group (UK) Ltd, Croydon, CR0 4YY
30/09/2026
02225699-0003